Professional Practice in Governance and Public Organizations

"Professional Practice in Governance and Public Organizations" offers cutting-edge insights and practical guidance for professionals in the areas of economics, politics, public policy and public administration, and those working at international organizations. The series features concise and accessible books on the latest developments in governance, organizational and political strategies, institutional policies, policy instruments, public management, and finance. Leadership and digitalization issues are a core topic throughout the series. All volumes are written by practitioners, experts and leading authorities from think tanks, non-governmental organizations, and public and international organizations. While the books are explicitly intended for professionals in the above-mentioned fields, students of economics, political science, public policy and public administration will also benefit from these practical guides for their future careers.

Ludovic Subran • Markus Zimmer

Investing in a Changing Climate

Navigating Challenges and Opportunities

Ludovic Subran
Allianz SE
München, Bayern, Germany

Markus Zimmer
Allianz SE
München, Bayern, Germany

ISSN 2731-9776 ISSN 2731-9784 (electronic)
Professional Practice in Governance and Public Organizations
ISBN 978-3-031-47171-1 ISBN 978-3-031-47172-8 (eBook)
https://doi.org/10.1007/978-3-031-47172-8

This Springer imprint is published by the registered company Springer Nature Switzerland AG
The registered company address is: Gewerbestrasse 11, 6330 Cham, Switzerland

Paper in this product is recyclable.

Acknowledgments

A book is seldom written solely by the authors. A number of highly dedicated colleagues from Allianz Research's network devoted countless hours not only to make many useful comments and valuable suggestions, but also to supporting us with careful, diligent search and data handling, and scientific reviewing, including Patrick Hoffmann, Arne Holzhausen, Stefan Landau, Anand Parmar, Verena Pichler, Simon Pichlmaier, Regina Reck, Georgina Rücker, Bridgette Stegmann, Lorenz Weimann, and Christian Wendlinger, which greatly improved the solidity of the messages the book wishes to convey. To all of them, a very special thanks. The unfailingly helpful disposition of Niko Chtouris and Parthiban Kannan at Springer Nature, in addition to the meticulous work of their editorial team, made sure that the entire publishing procedure ran remarkably smoothly and that the final product met the highest quality standards. Our sincerest thanks to each of them too.

Last but not least, a very special thanks to Julio Saavedra for his incisive comments and careful editing that turned the book's at times heavily technical nature into something not only much easier to digest, but even enjoyable to read.

Contents

About the Authors

Ludovic Subran is the Chief Economist and Head of Economic Research at Allianz SE, Europe's largest insurer and investor. A French national, he is a Member of the Council of Economic Advisors to the Prime Minister. Before joining Allianz, Subran worked for the French Ministry of Finance, the United Nations, and the World Bank. He also taught economics at HEC Business School and Sciences Po in Paris. Subran is a Young Global Leader of the World Economic Forum, a Millennium Fellow of the Atlantic Council, and a David Rockefeller Fellow of the Trilateral Commission.

Markus Zimmer is the Senior Economist for Sustainability at Allianz Research, where he heads all climate economic research activities. He links the business activities of Allianz to the global climate science community, fostering a fruitful exchange for both sides. Previously, he worked at the Energy, Climate and Resources Center of Germany's Ifo Institute at the University of Munich, where he was a researcher, policy consultant, and lecturer. He contributes with his expertise to the United Nations-convened Net-Zero Asset Owner Alliance and the Glasgow Financial Alliance for Net Zero.

1

Introduction

The marching orders are clear: avert the threat posed by climate change by limiting global warming to 1.5°C. Or was it to achieve carbon neutrality by 2050? Or reaching *Net Zero*, to use its meme-friendlier moniker?

As it turns out, the two goals—1.5°C maximum warming and Net Zero—are not one and the same. In fact, reaching Net Zero by 2050 will not be sufficient to stay below 1.5°C. To muddle things further, there are several Net Zeroes, some to be achieved well before 2050, and even a couple that ought to actually go *below* zero, into negative-emissions territory.

But if one could be excused for feeling somewhat mystified by the profusion of goals, the confusion only grows worse when it comes to how to reach them. The road to Net Zero—*any* net zero—is obscured by a thicket of subsidies, bans, emissions pricing, caps, offsets, levies, international accords, national goals, behavioural changes, and a bewildering array of technological fixes and geoengineering proposals, most of which are yet to be developed.

The one thing that is indisputably clear is that getting there will be costly. But, also indisputably clear, doing nothing would be far, far costlier.

Governments, as well as private and institutional investors, will therefore need to be very selective when it comes to deciding where to put their money—particularly now, when state and private purses alike are very tight in the wake of the covid pandemic, the energy crisis, ballooning public debt, high interest rates and galloping inflation. Climate action, clearly, must not only be effective, but also efficient, if only to make it easier to share the burden.

The original version of the chapter has been revised. A correction to this chapter can be found at https://doi.org/10.1007/978-3-031-47172-8_9

L. Subran, M. Zimmer, *Investing in a Changing Climate*, Professional Practice in Governance and Public Organizations, https://doi.org/10.1007/978-3-031-47172-8_1

Unfortunately, when sifting through the masses of data, reports, and hype, it is very difficult to find out what really merits further attention and, most of all, where the funds should go. A thorough review of the literature more often than not comes up empty in terms of clear, credible advice in this regard.

This book aims to correct that.

Based on insights gained by perusing reams of data and scientific papers, reviewing technological advances in a dozen critical industries, and even separating the wheat from the chaff in journalistic reports, the book lays out what the real state-of-the-art is on the various fronts and which approaches show the most promise—both for the climate and for investors. As a result, just like in a modern smartphone, the complexity is hidden and the ease-of-use a given.

But first, just to see how we humans got into this bind, it pays to take a brief look back. Already in 1896, Swedish scientist Svante Arrhenius published calculations showing that adding carbon dioxide to the atmosphere could warm the planet.[1] Given that humankind had been adding carbon dioxide at an increasing rate since the dawn of the Industrial Revolution, about a century earlier, it followed that humanity could over time change the climate on a global scale. The publication, however, was not intended as a warning, just as a noteworthy scientific possibility.

By 1960, in the wake of two seminal studies, scientists had started to worry that that scientific possibility might actually come to pass. One of the studies, published in 1957 by Roger Revelle,[2] had found that the ocean would not be able to absorb all of the carbon dioxide released by humanity's economic activities, leading eventually to a significant rise in carbon dioxide levels in the atmosphere. And, just as that study predicted, the second study, conducted three years later by Charles Keeling,[3] detected an annual rise in carbon dioxide levels in Earth's atmosphere.

The thinking, though, was that such potential changes would not have a noticeable impact until sometime in the twenty-first century, which seemed very far away. Notwithstanding the concern's rather nebulous nature, it caught the eye of the CIA, which in 1974 published a report on climatological change "as it pertains to intelligence problems".[4] While not necessarily focused on

---

[1] Weart, S. (2008). *The Discovery of Global Warming*. Harvard University Press.

[2] Roger Revelle, & Hans E. Suess (1957). *Carbon Dioxide Exchange Between Atmosphere and Ocean and the Question of an Increase of Atmospheric CO2 during the Past Decades*. Tellus, 9(1), 18–27. https://doi.org/10.3402/tellusa.v9i1.9075

[3] Charles D. Keeling (1960). *The Concentration and Isotopic Abundances of Carbon Dioxide in the Atmosphere*. Tellus, 12(2), 200–203. https://doi.org/10.3402/tellusa.v12i2.9366

[4] Central Intelligence Agency (1974). *A Study on Climatological Research as it pertains to Intelligence Problems*

rising temperatures, it warned of political unrest and mass migrations result-ing from "detrimental global climate change".

All of the above studies turned out to be spot on.

Some ten years after the CIA report, the influence of human activity on climate warming had already gone from theory to an established fact: every atom of carbon that we extract from the ground ends up eventually in the air as carbon dioxide. A large portion is absorbed by oceans and forests, but a significant amount remains in the atmosphere—for tens of thousands of years, warming the planet.

By now, the scientific community is by and large united in pleading for actions to be undertaken to forestall further warming. Graphic evidence abounds, from heat waves in Siberia and Canada—of all places—to devastat-ing forest fires in California and southern Europe; from deadly floods in China, Germany, Brazil, South Africa, USA and Pakistan, to name but a few, to the worst droughts in decades in every continent.

And all that from a temperature rise of less than 1.1 °Celsius so far. Even the most diehard sceptics find it now difficult to continue arguing that this is simply the weather doing its thing, as it has always done in the past.

But this nearly universal acknowledgement of the problem has not led to a united, coordinated, globe-spanning effort to combat climate change and avert the catastrophe. Even within countries a united front has often failed to emerge. The case of the United States is particularly glaring, with climate-change attitudes largely determined by which side of the political divide one sits on. Even the Supreme Court is taking sides—and not always to the ben-efit of the climate.

On the international front, it is not only poorer countries complaining that they did not cause the problem in the first place, and that they therefore deserve massive support to decarbonize their economies, a commitment that richer nations are loth to make. Far more worrying is the fact that four of the five largest emitters are at loggerheads with each other: the US and the EU (second- and fourth-largest emitters) are not talking to Russia (fifth-largest), while the US and China (the latter being the largest emitter overall) are locked in a geopolitical contest that will take years to resolve. India (third-largest), in turn, has increased its emissions by more than 300% since 1990. One positive glimmer, however, is that deforestation of the Amazon in Brazil (seventh-largest emitter) declined by a third in the first six months of Luiz Inácio Lula da Silva being in office, thus giving a fillip to a key global carbon sink.

This makes it even more important to quickly develop and deploy measures to at least slow down the increase of greenhouse-gas emissions, and ideally reverse it. The first step in this endeavour is, naturally, to identify and decide which measures are the most promising. That is where this book comes in.

To ease the way through the thicket, the book keeps to the science to assess the potential of both emission-reducing and mitigating approaches, grouping the analyses according to the main contributing sectors: energy, utilities, transportation, industry, buildings, and agriculture and forestry. The goal, however, is not to dissect and explain the intricacies, especially the engineering ones, of each approach, but to provide a sound foundation for investors to decide which ones to back.

The book is structured along the energy flow from primary energy production to secondary energy—i.e., energy transformation—to final energy use.

Chapter 2 tackles energy, which not only accounts for nearly three-quarters of the EU's total greenhouse-gas emissions, but also underlies every other sector. Whatever the improvements in the other sectors' carbon footprint, they will not amount to much in the overall quest for less-than-1.5 °C if the energy sector as a whole, from mining, extracting, refining and distribution, does not accomplish a major transition to low- or zero-carbon alternatives. This chapter also includes waste in its role as an energy source.

Chapter 3 examines the role of utilities. Electrification is slated to play an increasingly pivotal role in decarbonizing the EU's economy: already one of the largest energy carriers, it is projected to grow significantly as transport and industry are electrified.

Chapter 4 deals with another key player: Transportation. In all its forms— road, rail, sea and air—transportation is a major energy consumer and, with it, a large source of emissions. It is also vital for the smooth functioning of every economy. Now, constantly on the public spotlight and policymakers' sights, it is on the verge of a fundamental transformation that will open opportunities for investors.

Chapter 5 reviews Manufacturing, a sector with less visible pockets of high energy consumption and emissions which, taken together, play a significant role as targets and drivers of emission reductions. Included are glassmaking and ceramics, textiles and leather, pulp and paper, machinery, carmaking, and electrical equipment—as well as energy-intensive and greenhouse-gas-heavy other industries, such as cement, steel, aluminium, chemicals and so on. Success in reducing their carbon footprint and the creation of new materials with better energy-conservation properties would not only be a great accomplishment on its own, but would also contribute significantly to making many other areas of economic activity more energy-efficient.

Chapter 6 addresses Buildings. With 40% of the EU's energy consumption and more than one-third of energy-related GHG emissions, buildings deserve much more attention, and much more stringent emissions-reduction goals, than currently granted. Buildings, furthermore, tend to stick around for much longer than, for instance, any means of transportation. This means getting them right at the outset, including not only their overall energy efficiency but also making better use of them as carbon sinks, will make a more significant contribution to keeping global warming below 1.5 °C.

Chapter 7 delves into Forestry, Agriculture, Food Chain, and Land-Use. The importance of agriculture and fisheries cannot be overstated, in terms of securing food in a domestic context and beyond. But it is also a significant source of GHG emissions. Forestry, in turn, is both a carbon emitter and a carbon sink, giving it a particularly pivotal role. Land-use change, likewise, can play a positive or a negative role in combatting climate change. This chapter and the previous one address thus the role of carbon sinks in addition to just focusing on carbon sources.

Chapter 8 spotlights Africa, a continent that is simultaneously one of the worst affected by climate change, the most likely to see its greenhouse gas emissions increase—and the one with the greatest potential for solving the West's, and the world's, energy transition and economic growth conundrum.

Welcome to our concise guide to the state of the battle against global warming, which we are confident will help readers navigate the challenges and opportunities, financial and otherwise, of striving for a carbon-neutral future.

Climate Scenarios

This book features four different climate scenarios. These are not forecasts, but projections in which economic development and activity are consistent with a given global warming level. For the scenarios with higher temperature increases, the period after 2050—not featured here—is of particular importance as well.

- 1.5 °C Net Zero: Global warming is kept below 1.5 °C with a 50% probability. The carbon budget for additional CO_2 emissions past 2020 is limited to 500Gt, and net-zero emissions, whereby "negative emissions" through carbon capture and storage (CCS) offset the remaining residual emissions, are globally reached by 2050 at the latest. This scenario represents the level of ambition that is necessary to avoid surpassing thresholds that drastically increase the likelihood of triggering tipping points. Once triggered, these tipping points induce cascade effects that can result in unforeseeable and catastrophic damages.

- Below 2 °C: Global warming is kept below 1.8 °C with a 50% probability, and below 1.9 °C with a 67% probability. The carbon budget for additional CO_2 emissions past 2020 is limited to 1000Gt. This is the global warming level in which climate-change-induced damages exceed the mitigation costs, even without triggering any tipping points for global catastrophic events.
- NDCs: 'Nationally Determined Contributions', the climate commitments submitted by countries in the process of international climate negotiations, i.e., countries' current climate-related policy pledges. In this scenario, global warming is kept below 2.6 °C with a 50% probability, and below 2.9 °C with a 67% probability. The carbon budget for additional CO_2 emissions past 2020 is limited to around 2500Gt.
- Current Policy: No additional climate policy is enforced compared to the 2020 state of climate policy. It would thus represent a step back from the current policy path. In this scenario, global warming is kept below 3.0 °C with a 50% probability, and below 4.1 °C with a 67% probability. To stay within this waring range, the carbon budget for additional CO_2 emissions past 2020 around 4300Gt. This scenario will likely trigger some tipping points. It could, for instance, lead to the collapse of the Gulf Stream or cause the Atlantic Meridional Overturning Circulation (AMOC) to run amok, with unforeseeable global consequences and damages.

2

Watts Next?
The Energy Pathway

Energy is the currency of the Universe. It is also the coin for all transactions in our planet's life processes, and it powers our economies and fuels our prosperity. Unfortunately, the way we currently gain and expend this energy threatens to soon make our planet inimical to our kind of life.

The lion's share of the energy we use comes from the energy stored in chemical compounds containing fossil carbon, such as oil, natural gas, and coal. Most of the carbon that we extract from the ground ends up in the atmosphere, in the form of carbon dioxide. A large proportion of it is subsequently absorbed by oceans and forests, but the rest floats in the air for tens of thousands of years, trapping the heat that is now warming our planet. Carbon dioxide is the main atmosphere-warming compound among the greenhouse gases (GHG).

In recognition of this, the European Union agreed in December 2020 to tighten its decarbonization targets to a minimum reduction of 55% in GHG emissions by 2030 (previously 40%), compared to 1990 levels, and to attain net-zero status by 2050. The publication of the European Climate Law in July 2021 then made it legally binding for member states to meet these targets. In the EU periphery, the UK went even a step farther, aiming to cut emissions by 68% by 2030 and 78% by 2035.[1]

Nearly three-quarters of the EU's total GHG emissions come from the production and use of energy, making decarbonization of the entire energy

---

[1] Climate Change Committee (2020). *The Sixth Carbon Budget - The UK's path to Net Zero.*

L. Subran, M. Zimmer, *Investing in a Changing Climate*, Professional Practice in Governance and Public Organizations, https://doi.org/10.1007/978-3-031-47172-8_2

system crucial to stop global warming. To work towards this ultimate goal, a reduction in energy demand is necessary, as capacities for sustainable alternatives are limited. As a result, recent revision of the EU Energy Efficiency Directive set new targets, aiming to reduce primary energy consumption by 41.6% and final energy consumption by 38%, which would allow for a decrease of the energy system's GHG emissions by the targeted 55% by 2030.[2]

But is the energy system turning green quickly enough? The short answer is that while the fossil-fuel share in the energy mix is declining, fossil fuels will continue to play a key role for several decades yet. The Russian war against Ukraine, furthermore, has grayed out the greening prospects considerably, with even coal making a comeback of sorts.

The two most common measurements of energy consumption are Gross Energy Consumption (GEC), which includes the energy expended to extract, transform, and distribute energy, as well as all non-energy use of oil, gas, and coal; and Final Energy Consumption (FEC),[3] which is what end-users actually consume for energy purposes.

In the EU, GEC amounted to 1421 million tonnes of oil equivalent (Mtoe) in 2021, while FEC clocked 968 Mtoe. Of this latter amount, oil accounted for 37%, natural gas for 22%, and electricity for 23% (Figs. 2.1 and 2.2).[4]

Interestingly, as the energy transformation takes hold, GEC will *increase* slightly in 2050, after having decreased continuously until 2030 (Fig. 2.2). The reason is the expected increase in electricity consumption not only to power the millions of electric vehicles that EU countries would like to see replace internal-combustion types by then, but also to produce so-called e-fuels, such as green hydrogen and other gases and liquids that are expected to fuel transportation and industrial processes.

The EU 27's gross energy production is also undergoing a fundamental transition. Gross energy production can be much larger than final energy consumption, since, for instance, the use of electricity to produce e-fuels suffers from significant conversion losses. Thus, while final energy consumption decreases strongly from 2030 to 2050, gross energy production increases over the same period (Fig. 2.1). In 2021, renewable energy already accounted for 41% of the EU's gross energy production, driven by the rising deployment of solar, wind and bioenergy facilities (66% increase since 2009).[5] At the same

---

[2] Council of the EU (Press release, 07/2025). *Council adopts energy efficiency directive* and point (27), page 17 of the legislation *Energy efficiency and amending Regulation (EU) 2023/955 (recast)*.

[3] Also called Final Energy Demand.

[4] Eurostat (Update 04/2023). *Complete energy balances (NRG_BAL_C)*.

[5] European Commission (2021). *EU Energy in Figures – Statistical Pocketbook 2021*.

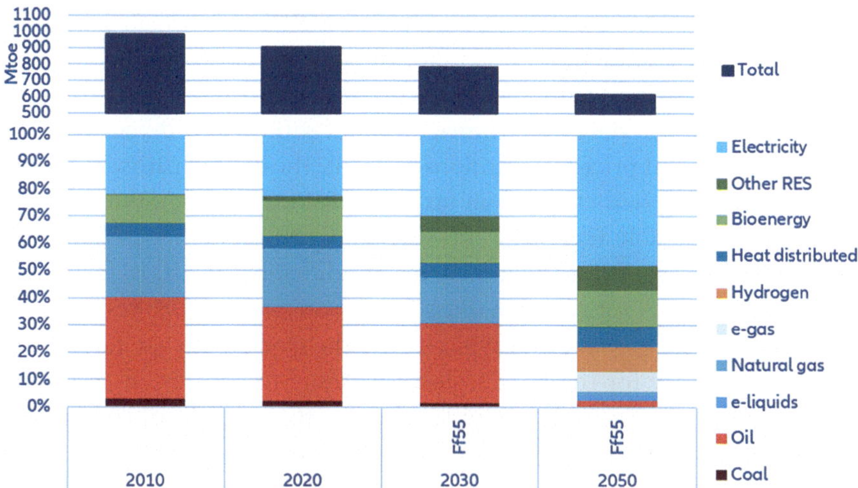

Fig. 2.1   EU final energy consumption. Sources: Allianz Research, EU Commission 2030 Climate Target Plan and JRC GECO

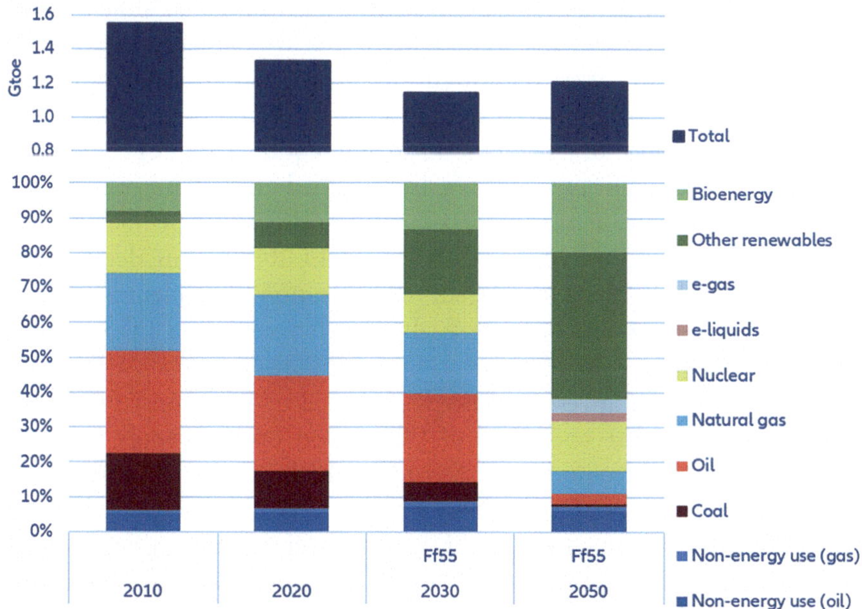

Fig. 2.2   EU gross energy production. Sources: Allianz Research, EU Commission 2030 Climate Target Plan and JRC GECO

time, fossil fuels show a declining trend, with a 39% decrease in solid fossil fuel production, a reduction of 42% for oil and petroleum products, and a 63% decline for natural gas. Nuclear energy also declined (−12%).[6]

Evidently, a single policy instrument will not suffice, as different products will require different price interventions to enable the desired pathway, including carbon pricing (either a carbon tax or Emissions Trading System, ETS), mandatory sustainable-fuel blend-ins, investment subsidies (focused on capital expenditures, CapEx), carbon contracts for difference (CCfDs, focused on operating expenses, OpEx) and markets for carbon removal. While oil and gas companies operate globally and typically use the global pathway for orientation, it is necessary for the EU's regional pathway to aim for an even more ambitious target to stay within the EU's share of the total global carbon budget.[7] A limited carbon budget also highlights the necessity for integrating carbon removal strategies into the energy companies' transition pathways, which would also help to achieve negative GHG intensity in the EU energy consumption product mix.

The trajectories shown in (Fig. 2.3) indicate that although the EU currently has a head start compared to the global pathway, the proposed annual reductions are not large enough, and will not occur quickly enough, to keep global warming below 1.5 °C. With the current targets, a 4-year implementation gap will open up from 2030 between a 1.5 °C-consistent and the proposed Ff55 pathway. Closing this implementation gap will require a total of EUR480bn of additional investment *per year* to be front-loaded until 2030, consisting of EUR80bn in supply-side investments (power grids, power plants, and production and distribution of new fuels), and EUR400bn in demand-side investments (such as in the industrial, transportation or residential sectors). This would come on top of what is already envisaged in a Ff55 scenario (Table 2.1).

The estimated total investment needs projected by Ff55 consist of annual averages between 2021 and 2030 of EUR120bn on the supply side, and EUR920bn on the demand side. According to the projection, investment volumes should increase further from 2031 to 2050, with expected annual average supply-side investments of EUR197bn and demand-side investment (including transportation) of EUR998bn. The composition of these investments remains practically unchanged.

---

[6] Eurostat (Update 04/2023). *Complete energy balances (NRG_BAL_C).*

[7] There are different concepts to determine the EU's share typically derived either from: (1) fairness considerations, (2) optimizing the transition in a cost-efficient manner, and (3) a technology perspective of working within feasible technological boundaries. The NGFS pathways mentioned in this book fall within the second category, while the OECM pathways tend to fall within the third category.

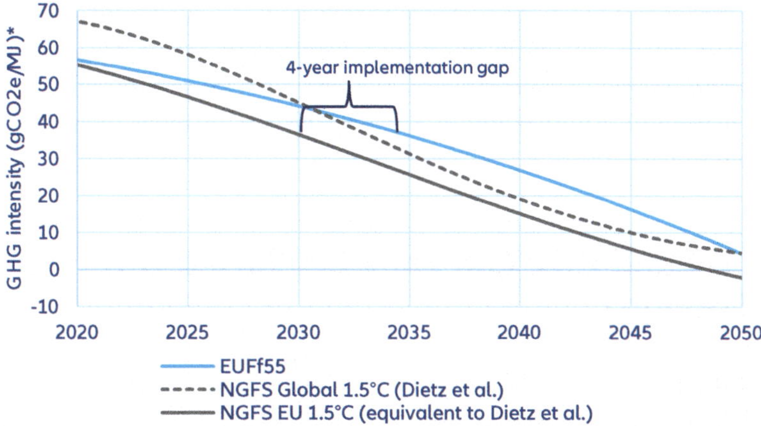

Fig. 2.3 GHG intensity of indicative pathways [EU Ff55 includes Scope 3 emissions down the value chain (category 11), for gaseous, liquid, and solid energy carriers, while scope 1 + 2 are only available for refineries. For this reason, scope 1 emissions prior to refineries are missing]. Source: Allianz Research, Dietz et al., EC 2030 Climate Target Plan. Pathway methodology in Dietz et al. (2021a, 2021b). NGFS = Network for Greening the Financial System, * gCO2e/MJ is grams of CO_2 equivalent per megajoule

Table 2.1 Estimated investment needs for the EU energy system as per Ff55

Category	Average annual investment in billion EUR	
	2021–2030	2031 2050
In Power grid	58.2	80.9
In Power plants	56.2	88.5
In boilers	3.8	1.3
In production and distribution of new fuels	1.4	26.6
Total supply side	119.9	197.3
Industrial sector	20.3	14.4
Residential sector	190.0	174.4
Tertiary sector	87.7	80.7
Transport sector	621.8	728.2
Total demand side	919.8	997.7
Total energy system	1039.7	1195.0

Sources: European Commission [European Commission (SWD/2020/176 final). *Stepping up Europe's 2030 climate ambition – Investing in a climate-neutral future for the benefit of our people*], Allianz Research

To put these investment volumes into perspective, a Ff55 future would require about 9% more investment than what is currently planned for 2021–2030, increasing to around 20% more investment from 2031 to 2050. However, as Fig. 2.3 shows, this would not be enough to achieve the goal of staying below 1.5 °C warming. Closing the implementation gap between the

Ff55 pathway and a 1.5 °C pathway would require investments to be 84% higher than what is currently envisioned over the rest of this decade.

Reducing energy consumption is but one element in the quest towards carbon neutrality. Another one is the phasing-out of fossil fuels at a sufficiently rapid pace.

Coal, Oil and Gas: Can They Be Phased Out Fast Enough?

Solid Fuels

One of the biggest challenges for the energy system is phasing out combustion of thermal coal by 2030 in the power sector.[8] Coal is primarily used for electricity generation, and is also a critical heat source and reduction agent in steel production. Despite a 27.76% rise in electricity production between 1990 and 2021, the power sector's GHG emissions dropped by 28.94% thanks to a steady replacement of coal with cleaner fuels, as well as to an increase in electricity generated from renewables.[9]

All Ff55 policy scenarios in the European Commission's assessment state that electricity generation from coal must be completely phased out by 2030 to reach an overall reduction of GHG emissions of 55%. This appears increasingly unlikely. Solid fossil fuels still account for at least 25% of the total energy mix of Bulgaria, the Czech Republic and Poland, with Poland, Germany and the Czech Republic taking the EU's top spot as producers of fossil fuels in 2021, totaling 80.06 Mtoe. Coal imports, which were declining in 2021, saw nearly one-third of the total go to Germany (32%) followed by Poland (12%), France (8%), the Netherlands (7%) and Italy (8%). Hard coal was chiefly imported from Russia (55 Mtoe), the United States (21 Mtoe), Australia (16 Mtoe), Colombia (10 Mtoe) and South Africa (3 Mtoe).[10]

On a positive note, most EU member states and the UK plan to phase out coal by 2030. Germany had initially committed to phasing it out by 2038, but its new coalition government has now announced that phase-out should occur "ideally" by 2030—a step in the right direction, but still lacking full

---

[8] For details on the coal exit, see Chap. 2 Utilities.

[9] Joint Research Center (JRC) (2020). *Towards net-zero emissions in the EU energy system by 2050.*

[10] European Commission (2021). *EU energy in figures - Statistical Pocketbook 2021.*

commitment. Meanwhile, the remaining Coal-5 countries (Poland, Bulgaria, the Czech Republic, Romania, and Slovenia) have not yet committed to phasing out coal by 2030. If the EU would like to completely phase out coal for electricity generation by 2030, approximately 100 GW of additional wind and solar, as well as 15 GW of natural gas power plants, would be needed to replace it.[11]

In the Net Zero Scenario of its 2021 World Energy Outlook, the International Energy Agency (IEA) stated that coal use must drop by 90% globally from 2020 levels by 2050. Additionally, 80% of the coal-powered energy generation facilities should incorporate carbon capture and storage (CCS) technology. The remaining coal in the industrial system is not used for electricity production, but rather for the chemical and iron/steel industries.

After coal, the next solid fuel of importance is solid bioenergy carriers. Consisting of biological feedstocks, such as wood, food waste, and agriculture residues, it falls under two categories: the traditional sort, used for cooking or charcoal production, and the modern sort, used to produce liquid or gaseous biofuels like ethanol for gasoline and biodiesel for distillates. In the EU, modern bioenergy is especially important in the production of biodiesel or sustainable aviation fuel.

Oil

The EU is still heavily depending on oil, which accounted for 35% of final energy demand in 2021. According to the Ff55-proposed scenario, oil is expected to remain a significant contributor until the end of this decade, when it is expected to decrease slightly, to 29% of final energy demand. By 2050, however, the situation is projected to change, with oil dropping to only 2% of final energy demand.

In the transportation sector, dependence on fossil fuels is nearly absolute (Fig. 2.4), with air and maritime transportation at almost 100%, and road transportation at 93%. Oil products are expected to continue to dominate throughout this decade. However, by 2050, oil products will need to drop substantially, accounting for no more than 13% of all fuels consumed in transportation. The remaining oil products in the fuel mix are expected to be used primarily in aviation and maritime transportation, for which finding sustainable alternatives is more challenging.

---

[11] See also Chap. 2 Utilities.

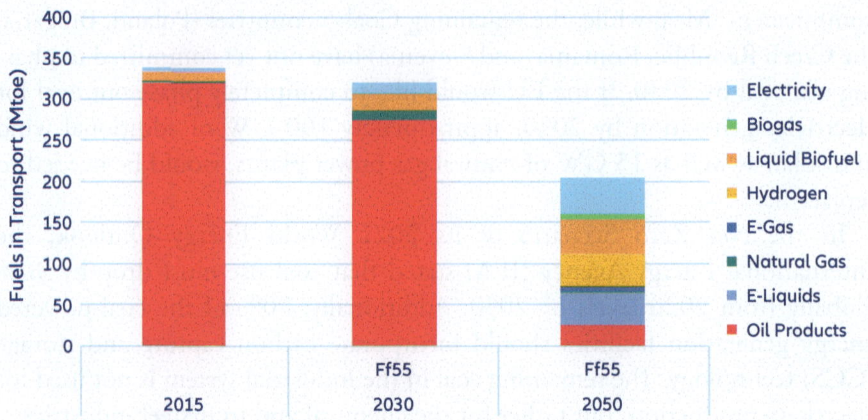

Fig. 2.4 Forecast of the fuel mix in the transportation sector. Sources: EU Commission 2030 Target Plan and JRC GECO

The EU-27's primary production of crude oil reached a record low of 17.6 Mtoe in 2021, with the leading producers being Italy (4.8 Mtoe), Denmark (3.3 Mtoe) and Romania (3.0). The EU's net import of crude oil was 444Mtoe in 2021, largely coming from Russia (112 Mtoe), Norway (44 Mtoe), the US (37 Mtoe), Kazakhstan (36 Mtoe), and Libya (36 Mtoe). The top oil importers were the Germany (18%), Italy (13%). Spain (13%), the Netherlands (12%) and France (8%). The flow of imports is changing rapidly. Imports from Russia and Saudi Arabia had already been decreasing before Russia invaded Ukraine, while imports from Iraq, Nigeria, and Kazakhstan increased. With the bans on seaborne crude oil (December 5, 2022) and on petroleum products (February 5, 2023), oil imports from Russia decreased by over 90% by March 2023 compared to 2019 levels, which additionally reshaped the flow of oil imports into the EU.

Gaseous Fuels

Gas may play an increasingly prominent role as rising carbon prices promote the switch in power generation from coal to gas. Before the Russia-Ukraine war, natural gas use was projected to decrease by only 13% until 2030, compared to 2015 levels. However, the stop of gas exports from Russia has forced the EU to accelerate the pace of replacement, switching initially to liquified natural gas (LNG) imports and later to green hydrogen, e-gas (gaseous fuels obtained using electricity) and biogases. Projections before the war saw hydrogen, e-gas and biogas overtaking natural gas as the main gaseous fuel, with the natural gas share dropping from 93% in 2030 to 32% by 2050 (Fig. 2.5). Needless to say, the replacement of natural gas with (green) hydrogen, e-gas and biogas offers significant CO_2 emissions reduction potential.

The changes in consumption rates of natural gas will also differ by sector (Fig. 2.6) as the power-generation sector is expected to increase its gaseous fuel consumption, from 30% in 2015 to 40% in 2050. The largest reductions in consumption by 2050, compared to 2015 levels, will come from the residential, services, and agriculture sectors (−23 percentage points), followed by industry (−10 pp).

For buildings, the most important single energy use of gas is for space heating and cooling. Over time, the fuel mix for buildings is expected to substantially shift to electricity, while fossil fuel consumption, especially natural gas, will fall. The main drivers of this shift will be higher carbon prices, increasing

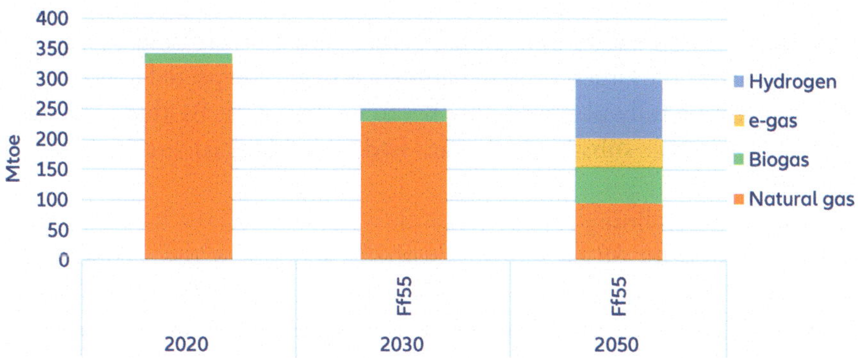

Fig. 2.5 Consumption of gaseous fuels per gas type. Source: Allianz Research, EU Commission 2030 Target Plan and JRC GECO

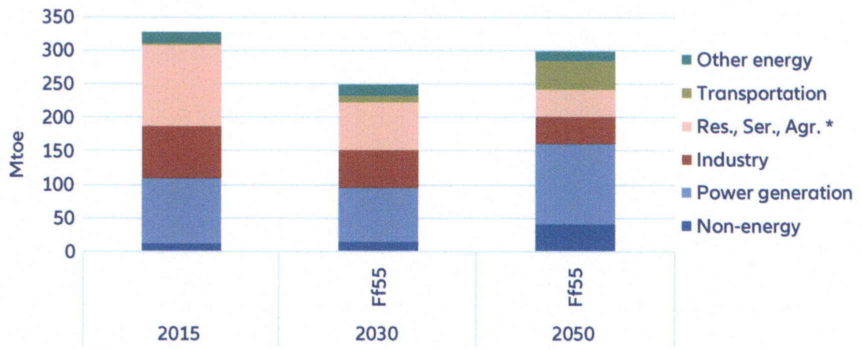

Fig. 2.6 Consumption of gaseous fuels per sector. Sources: EU Commission JRC GECO, * Residential, services, agriculture

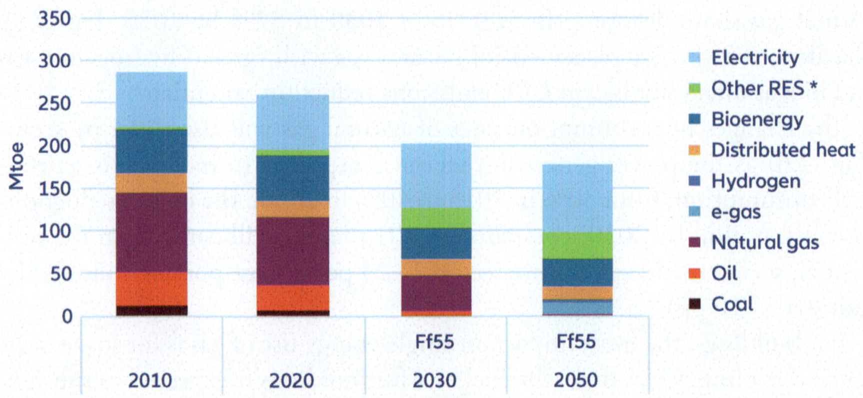

Fig. 2.7 Energy demand in residential buildings. Sources: EU Commission JRC GECO, Allianz Research, * Other renewable energy sources

deployment of renewables in heating/cooling, and an increasing adoption of heat pumps. But the largest drop in natural gas usage is expected to occur after 2030 and will be fully observed only by 2050 (Fig. 2.7). By this time, more sustainable gaseous substitutes (e-gas, hydrogen, biogas) are expected to dominate.

As a result of declining production and rising demand, the EU has become the largest importer of gas globally, with natural gas as the second-largest imported energy product after oil. The EU produced 39 Mtoe of natural gas in 2022, primarily in the Netherlands (14 Mtoe), Romania (8Mtoe) and Germany (4). Net imports (Fig. 2.8) reached 303 Mtoe in 2020, coming mainly from Russia (137 Mtoe), Norway (60 Mtoe), Algeria (27 Mtoe) and Qatar (15 Mtoe). Largest importers were Germany (24%), Italy (20%), France (12%), Spain (10%) and Belgium (6%).[12]

---

[12] European Commission (2021). *EU Energy in Figures – Statistical Pocketbook 2021.*

Fig. 2.8   Origin of energy imports. Sources: EUROSTAT, Allianz Research

Energy Dependency and Other Associated Risks

The energy dependency rate measures the reliance of an economy on imports to meet its energy demand. In 2020, the EU's import dependency for all fuels dropped to 57.5%, from its peak of 60.5% in 2019[13] (Fig. 2.9), which meant that most of the gross inland energy consumption was accounted for by net imports.[14] It goes without saying that such high dependency creates geopolitical vulnerabilities, as the effects of the Russia–Ukraine war and the ensuing energy crisis in Europe has shown.

Overall fuel dependency is expected to decrease over time, however, to 17% by 2050 (Fig. 2.2), thanks to a reduction in fossil fuel imports and an increase in domestic renewable production. By that time, coal imports are expected to nearly disappear, while natural gas and oil imports are projected

[13] Eurostat (2022). *Archive: EU energy mix and import dependency.*
[14] European Commission (2021). *EU Energy in Figures – Statistical Pocketbook 2021.*

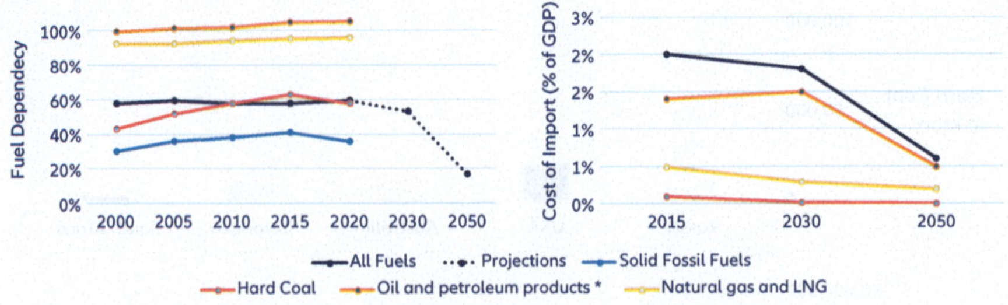

Fig. 2.9 Import dependency and cost of imports (as % of GDP) by fuel type [Imports from extra-EU]. Sources: Allianz Research, European Commission 2030 climate target plan, EU energy in figures—Statistical pocketbook 2022, EUROSTAT

to decrease by 62% and 78%, respectively, compared to 2015. Thus, transitioning from fossil fuels to renewable energy in the energy production mix will not only reduce emissions, but also increase energy independence, security and supply resilience.[15]

In line with decreasing imports, the cost of imports was also expected to decrease, until the current energy crisis hit. For all fuels, the cost of imports (as a percentage of GDP) was expected to decrease from 2% in 2015 to 1.8% in 2030 and to 0.6% in 2050. This decrease was expected to result in cumulative savings in net energy imports of between EUR83bn and EUR133bn, compared to the EU baseline scenario. The Ukraine war has upended these projections, with no clarity expected at least until after 2024.

But there are more risks associated with fossil fuels than the ones derived from geopolitical shifts. From a climate perspective, methane leakage poses a significant threat. Methane (CH_4) is a more potent GHG than CO_2 and ranks second to it in the overall contribution to climate change. The global warming potential of one ton of methane is assumed to be equivalent to 29.8 tons of carbon dioxide over a 100-year timeframe.[16] The EU 2030 climate target plan impact assessment suggests that CH_4 will retain its status as the dominant non-CO_2 GHG in the EU. Due to its substantial climate impact, the EU and the US have set the goal of reducing anthropogenic methane emissions by 30% by 2030 compared to 2020 levels.[17]

The EU accounts for 5% of global methane emissions. Out of the EU's total anthropogenic methane emissions, 19% stem from energy production

---

[15] The issue of intermittent energy sources on resilience will be addressed in *Chap. 2 Utilities*.

[16] IPCC (2021). *Sixth Assessment Report*.

[17] European Commission (2022). *Fact Sheet - Global Methane Pledge: From Moment to Momentum*.

and use. Accidental methane leaks are a by-product of fossil fuel production and distribution. Estimates suggest that 54% of energy-related methane emissions are fugitive emissions from the oil and gas sector, while 34% come from the coal sector. The EU's climate target plan impact assessment shows that the most cost-effective methane emission savings can be achieved in the energy sector.

The line of action of the European Commission (EC) for emission reductions in the energy sector is to support voluntary initiatives by fostering the widespread implementation of a measurement and reporting framework covering oil and gas upstream companies, under the Oil and Gas Methane Partnership (OGMP). The United Nations Environment Programme (UNEP) and the Climate and Clean Air Coalition are cooperating to extend the framework to gas midstream and downstream. Alongside these endeavors, legislation is also being drafted to reinforce these actions through compulsory measurement, reporting and verification of all energy-related methane emissions, which builds on the OGMP methodology. The EC also encourages companies in the oil, gas and coal industries to prepare leak detection and repair (LDAR) programs in preparation of forthcoming legislative proposals that would make them mandatory.[18]

Achieving emissions savings in the oil and gas industry is highly feasible, since at least one-third of reductions is possible at no net cost to the industry. Reducing methane emissions from venting and flaring, LDAR in natural gas, coal, and oil production, transmission, and combustion, promise sizable benefits in economic, environmental, and social terms.

The Role of Individual Corporations

No oil and gas company will be able to sidestep the transition towards clean energy. The entire industry needs to have a plan of action to weather the transition, which should be in line with the anticipated decrease in demand. A key aspect to the decarbonization pathway is maintaining an equilibrium between phasing-out of fossil fuels and phasing-in of renewables. This can best be achieved by looking at the demand and supply side of fossil fuels simultaneously: both supply and demand need to be wound down in tandem, without creating dangerous mismatches. Focusing solely on demand (i.e., through

---

[18] European Commission (COM/2020/663 final). *EU Strategy to reduce methane emissions*. See also US Environmental Protection Agency (2007). *Leak Detection and Repair – A Best Practices Guide*.

higher carbon prices) could lead to a situation where the supply side might be tempted to slow down the transition, be it by lowering prices or (covert and overt) lobbying activities. Investors can play a key role in incentivizing the supply-demand balance and helping to accelerate the transition.

The capacity of companies to achieve an environmentally and socially just transition is gaining prominence among the criteria considered by investors. This includes looking at the dialogue a company has with such stakeholders as trade unions and local communities, its track record of successful transformation, and corporate responsibility actions. More companies are applying frameworks for sustainable investment, which, from a risk perspective, is a form of future-proofing. The finance industry generally recognizes that the oil and gas sector will continue to play an important role in satisfying the energy needs of the global economy, and therefore aims to support the sector as an insurer and investor when it takes measures to drive the transition.

Role of Private Equity Firms in Transformation

While institutional investors push for the transition of the energy sector, private equity seems to move in the opposite direction. Since 2010, private-equity firms have invested around USD1.1 trillion into energy assets. In 2020, these firms owned over 300 portfolio companies across the energy sector, with 80% in fossil fuel and only 20% in renewable assets.[19] Under public pressure to decarbonize portfolios, some public companies have sold off their climate-sensitive assets and private equity firms have stepped up to purchase, which simply shifts the operations (and emissions) from the spotlight to the shadows. A report by the Private Equity Stakeholder Project highlights several examples of how publicly traded oil majors have sold segments of their operations to private equity groups.[20] Rather than supporting climate solutions, this private investment is being funneled to sustaining, expanding, and dirtying fossil fuel energy, which is environmentally, socially—and arguably economically—unsustainable.

The current lack of transparency can be attributed to an absence of regulations that direct private equity firms to disclose their climate impacts. This

---

[19] Private Equity Stakeholder Project (2021). *Private Equity Propels the Climate Crisis* and Clean Air Task Force & Ceres (2022). *Benchmarking Methane and Other GHG Emissions Of Oil & Natural Gas Production in the United States.*

[20] The New York Times (2021). *Private equity funds, sensing profit in tumult, are propping up oil.*

could (and should) change in the future as the US Securities and Exchange Commission (SEC) investigates how climate disclosure requirements should be updated to include private equity firms as well.

When assessing a private company's emission reduction efforts, the scope of an emission is an important component to better understand the source of the emissions. Although CO_2 emissions arising from the burning of fossil fuels are well-known, GHG emissions resulting from the extraction, processing and transportation of oil and gas are often less scrutinized. In general, energy sector emissions can be categorized using the scopes defined by the Greenhouse Gas Protocol.[21]

Scope 1 consists of GHG emitted directly by the energy producer through, for example, combustion of fossil fuels, powering drilling equipment, refining, methane leaks and fuel transportation in company-owned vehicles.

Scope 2 accounts for the emissions emanating from the generation of electricity, heat, or cold purchased by the producer (self-production falls under scope 1).

Scope 3 emissions are other indirect GHG emissions. They cover about 15 categories, such as the extraction and production of materials, transportation, business travel, use of products sold (e. g. combustion of fuels sold) or waste disposal.

The IEA World Energy Model provides estimates for GHG emissions by tracking a barrel of oil or a cubic meter of gas from its production site to final consumption. According to this model, a barrel of oil's journey to its user emits 95 kg of CO_2. For gas, the emissions are around 100 kg CO_2eq per barrel of oil equivalent (boe).[22]

Providing such estimates accurately is complex, but on average the combined scope 1 and 2 emissions for oil and gas companies account for about 20% of the total life-cycle emissions (which also include scope 3 emissions) of oil, and for 25% in the case of natural gas. Thus, although Scope 3 emissions (the emissions resulting from the final consumer burning the fossil fuel) hog the limelight, given that they account for the largest share of total emissions, scope 1 and 2 emissions (from extracting and transporting oil and gas) are still significant sources of GHGs. There is, however, a wide range of emission intensities across different sources of production for both oil and natural gas, with methane leakage being the largest source of emissions on the journey from reservoir to consumer.

---

[21] World Business Council for Sustainable Development (WBCSD) & World Resource Institute (WRI) (2004). *The Greenhouse Gas Protocol: A Corporate Accounting and Reporting Standard.*

[22] International Energy Agency (2020). *The Oil and Gas Industry in Energy Transition.*

A Concrete Look at Oil and Gas Companies

The energy sector's fossil fuel-related enterprises are an amalgam of companies engaged in exploration, production, refining, marketing, storage, and transportation of coal, gas and oil along with other consumable fuels. The EU has 166,188 enterprises in the energy sector, employing 1.69 million people and generating EUR1.9Trn in turnover. Table 2.2 shows a breakdown of the number of selected types of enterprises across the energy sector.

As energy suppliers, the scope and timeframe under which fossil fuel companies plan to decarbonize is an important component in the energy sector transition. Their individual plans, in the aggregate, should delineate a joint transition that falls in line with a 1.5 °C maximum warming.

The collective emission reduction pathways can be approached in two ways: a sector-based approach using convergence, or an absolute-based approach using contraction (Fig. 2.10).

Under the first, the sector-based approach, where the carbon intensity for all companies converges towards the same 1.5 °C-compatible pathway, companies can decarbonize at different rates, but in the aggregate move towards a common level. This will result in meeting the 1.5 °C carbon budget only if companies are (production-weighted) equally distributed above and below the necessary or reference pathway. If, for example, more companies start out above the reference pathway, too much GHG would be produced until the reference level is reached.

Table 2.2 Number of energy sector enterprises in EU-27

Focus	Number of Enterprises (2019)	Trend since 2015
Mining of Coal and Lignite	198	−19%
Extraction of Crude Petroleum & Natural Gas	246	+10%
Extraction of Peat	910	−8%
Support Activities for Petroleum and Natural Gas extraction	1002	+15%
Manufacture of Coke and Refined Petroleum Products	841	−3%
Manufacture of gas	3823	NA
Production of electricity	133,975	+46%

Sources: European Commission [European Commission (SWD/2020/176 final). Stepping up Europe's 2030 climate ambition – Investing in a climate-neutral future for the benefit of our people. See also Eurostat (Update 03/2023). *Annual enterprise statistics by size class for special aggregates of activities (NACE Rev. 2) (SBS_SC_SCA_R2)*], Allianz Research

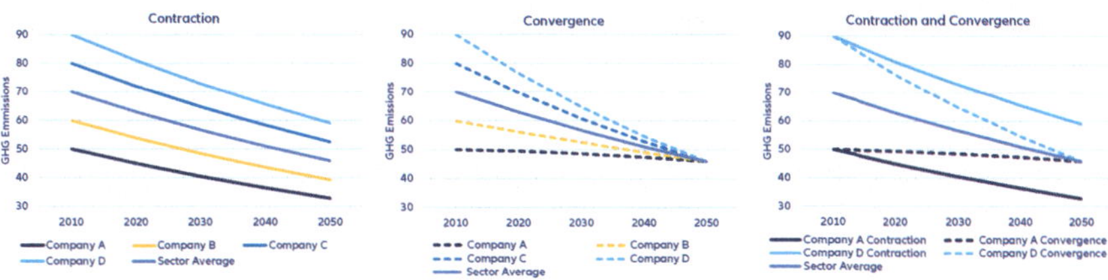

Fig. 2.10 Emission reduction pathway approaches: Contraction vs Convergence. Source: Allianz Research

Under the absolute-based approach, all companies reduce emissions by the same absolute percentage as in the reference pathway. This approach, too, suffers from the same distributional problems as the previous one, as most companies are currently far above the necessary pathway. The problem is exacerbated by self-selection: since companies are free to choose their preferred approach, those above the sector reference pathway will choose the contraction approach, while those below will benefit from choosing the convergence approach, thus further inflating the sector's carbon usage. Given that the oil and gas sector is on average above the reference pathway, it seems advisable to recommend the convergence approach as the standard procedure to stay aligned with the 1.5 °C target.[23]

Moreover, pathways also differ regionally. As depicted in Fig. 2.11, the regional Current Policies pathway for the EU GHG intensity is below the global average, meaning that the 1.5 °C pathway will even need to reach negative intensities through carbon dioxide removal (CDR). Demand intensity should be based on regional pathway values that are in line with the regional carbon budget. On the other hand, supply intensity should be based on the average global demand intensity, which needs to be consistent with the aggregate of the regional intensities. The EU pathways and high-temperature projections in the figure were derived in a similar manner as the methodology for the global pathways in Dietz et al. (2021a, 2021b).[24]

Using a general convergence approach, we developed pathways for selected oil and gas companies based on the regional and global pathways across varying warming scenarios (1.5 °C to 4 °C). The results, shown in Fig. 2.12, are calculated according to Dietz et al. (2021a, 2021b), NGFS and SBTi. The pathways describe how the GHG intensity of energy products sold should

---

[23] Note that both approaches can be equivalent, for instance in this example if an emission intensity of zero was to be reached in 2050.

[24] The pathways have different starting points in 2020, since the base year lies before 2020.

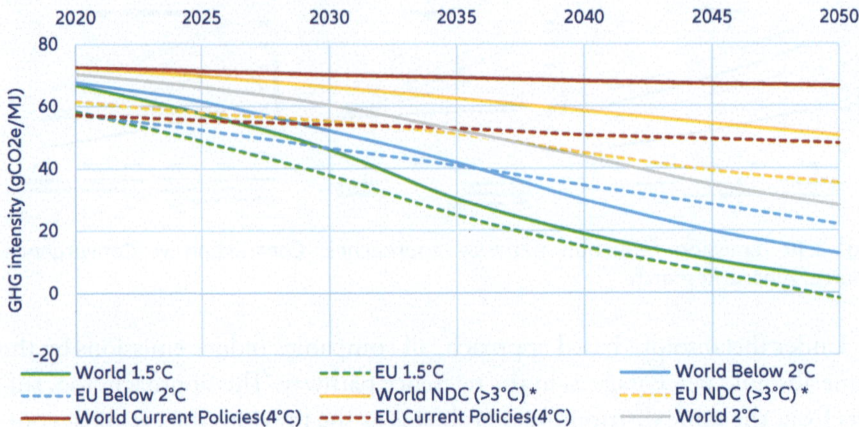

Fig. 2.11 Pathways for emission-intensity reduction, by climate scenario. Source: Allianz Research, * NDC refers to Nationally Determined Contributions

evolve under the various scenarios, and include scope 1, 2, and 3 emissions.[25] Scope 3 emissions are calculated applying emission factors to the products sold. To achieve better comparability and to obtain a more complete company list, a standardized reporting of the products sold would be required.[26]

---

[25] Base-year intensities are taken from Dietz et al. (*2021a, 2021b*) and are subject to confidence intervals that are included in the original publication. Dietz et al. also list the breakdown in scope 1, 2 and 3 emissions.

[26] For more details see Markus Zimmer et al. (2021). *Jostle the colossal fossil – A path to the energy sector transition.*

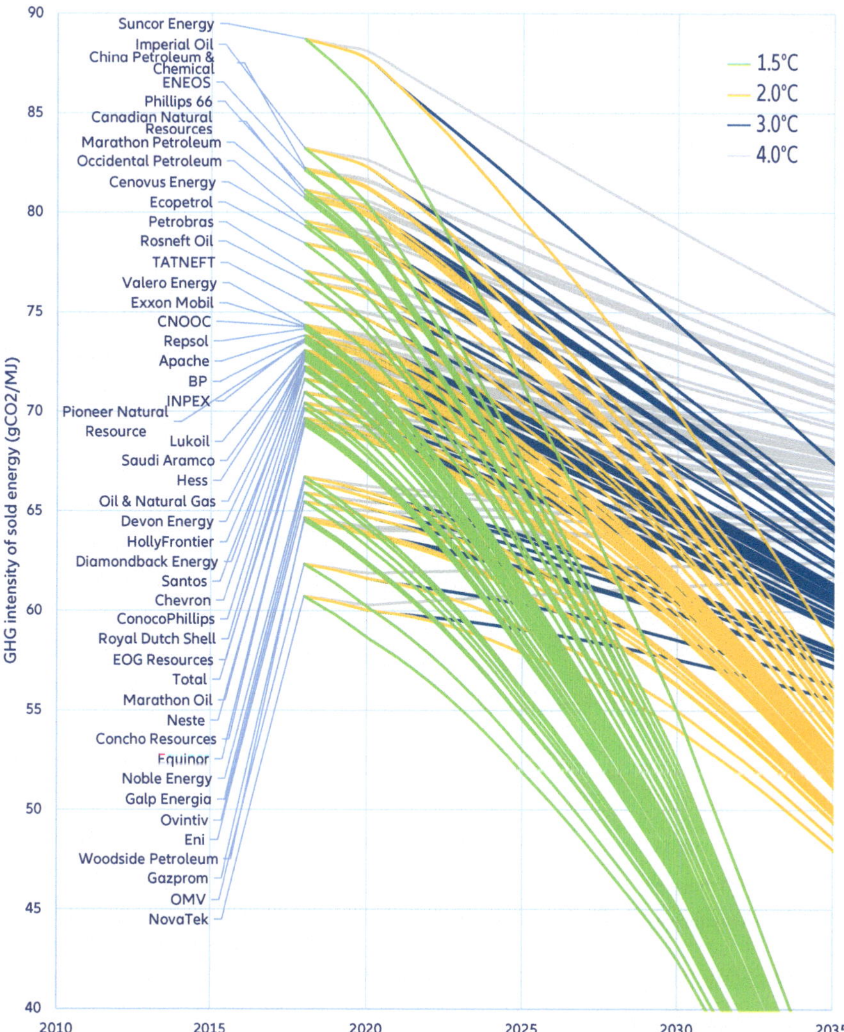

Fig. 2.12   1.5 °C, 2 °C, 3 °C and 4 °C convergence pathways of combined scope 1 + 2 + 3 GHG emission intensity of energy products for selected oil and gas companies. Source: Allianz Research, own calculations based on Dietz et al. (2021a, 2021b), SBTi and NGFS

Where Do We Go from Here?

Now, more than ever, the decisions and actions of private and public corporations will play an increasingly important role. It is, however, unrealistic to expect overnight action. Fortunately, there has never been a better time to ramp up investments in renewable energy alternatives. The cost of capital, which a project needs to recover to be profitable, is 15 percentage points lower for the average renewable energy project than for the average fossil fuel-based undertaking (Fig. 2.13). Finding financing for new fossil fuel projects is becoming more difficult due to the pressure that banks are facing to green up their portfolios. This pressure results in more expensive debt financing, which has recently increased the costs of capital for oil companies to over 20%.

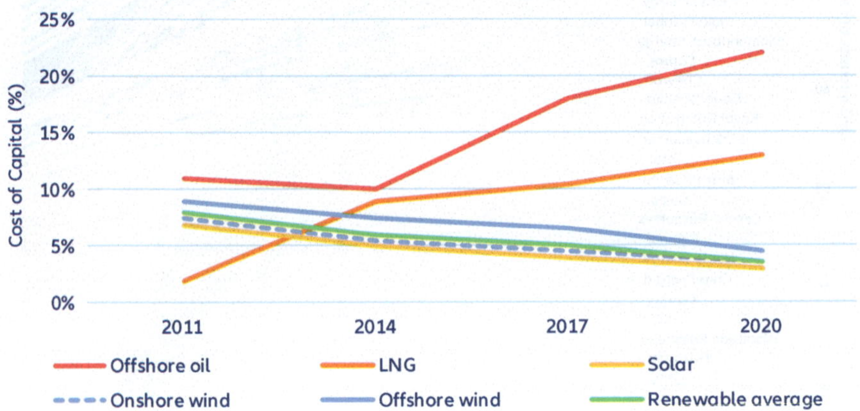

Fig. 2.13 Cost of Capital: Fossil Fuels vs Renewable Energy. Source: Allianz Research

H2 Can Do

For a colorless gas, hydrogen comes in a wide range of hues. It goes from the black sort, which is the dirtiest and made from coal, through its cousin the gray one, made from natural gas. Both produce large amounts of carbon dioxide. Then comes the blue variety, made in exactly the same way as the gray one, but in which the resulting CO_2 is captured and stored permanently underground. Next comes turquoise hydrogen, made from methane pyrolysis, in which heat separates the hydrogen and leaves solid carbon as a byproduct. Then, pink hydrogen, made using nuclear power. And, finally, green hydrogen, produced using renewable energy sources to split water molecules into hydrogen and oxygen. This last one is the darling of those advocating for a decarbonized future.

Unfortunately, green hydrogen accounts at present for only around 0.1% of the 90 million tons of hydrogen produced every year.

But this very low volume of the only version of a fuel that has impeccable green credentials makes it such a hot area for investment. Besides climate considerations, several other factors bolster the case for green hydrogen. For one, the current energy crisis has laid bare the folly of overly depending on unreliable energy suppliers. For another, the spike in fossil fuel prices, in particular that of natural gas, which makes gray hydrogen more expensive, has made green hydrogen all the more economically viable. Finally, and crucially, demand for green hydrogen is practically unlimited.

Therefore, it comes as no surprise that the EU has an ambitious vision of significantly raising the share of hydrogen in its energy mix, busily entering into partnerships with countries, such as the United Arab Emirates, Saudi Arabia and Chile, among others, that possess good green hydrogen-producing potential.

What makes hydrogen such a promising fuel?

Hydrogen has one of the highest energy densities by weight, second only to nuclear power, and three times the energy density (also by weight) than diesel or gasoline, while producing zero carbon emissions when combusted. To make its green credentials perfect, it needs to be produced using solely energy from renewable sources.

A further advantage of hydrogen is its versatility: it can be used as fuel for transportation, feedstock for industrial processes, as an energy carrier, for energy storage, or mixed with natural gas for building heating. It is particularly valuable to help decarbonize heavy road transport and energy-intensive industries.[27]

Much of the initial focus has turned to industrial applications, which include the steel and refining industries, as early markets for on-site hydrogen production.[28] The European Hydrogen Roadmap[29] projects that by 2050, hydrogen could provide approximately 2251 TWh of energy in the EU, while DNV[30] reckons

---

[27] See Chap. 3 Transportation for further details, including the role of further processing hydrogen to sustainable aviation fuels or to ammonia for shipping.

[28] See Chap. 4 Industry for more details including other industries like aluminum and cement.

[29] European Union (2016). *The European Hydrogen Roadmap - A sustainable pathway for the European energy transition.*

[30] DNV (2021). *Energy Transition Outlook.*

that hydrogen will supply 1250 TWh, with varying estimations on the breakdown by sector (Table 2.3). Based on the total final energy demand estimated in the Ff55 scenario (7152 TWh), hydrogen could meet between 17% and 31% of Europe's final energy demand.

To get there, however, production must be ramped up as quickly as possible, for which investment must start now. Although hydrogen demand is only expected to double until 2030 (before potentially increasing sevenfold by 2050), the market must be ready for its production, distribution, and deployment between 2030 and 2050.[31] The European Hydrogen Roadmap estimates that approximately EUR60bn in total investment is needed by 2030, with 40% of the investment share going towards setting up infrastructure and equipment for hydrogen production and distribution. The remaining investment would be funneled towards storage, buffering, and retail in transportation, buildings, and industry applications.

This provides investors with a range of green, profitable opportunities: the market could be worth anywhere from EUR85bn to EUR150bn by 2030, with some export potential as well. By 2050, the market could reach EUR820bn. Given the European industry's know-how, it could potentially capture 75–90% of the domestic market revenues. If this comes to pass, an estimated one million jobs could be added by 2030, rising to 5.4 million by 2050.

The cost competitiveness of hydrogen is still low today, but expected to increase slowly over time, making investment more attractive in the long term. Currently, natural gas-based hydrogen ("gray hydrogen") can be produced for approximately EUR1.5 per kg. This cost rises to EUR2 per kg if the CO_2 generated during production is captured and stored ("blue hydrogen"). In a high gas price environment, as observed in 2022, grey and blue hydrogen costs increase substantially, to EUR2.65 per kg for the former and EUR3 per kg for the latter. Production costs of green hydrogen in the EU currently average around EUR3.3–6.5 per kg.[32] Over time though, the levelized cost[33] of low-carbon hydrogen is expected to fall below EUR2 per kg by 2050.

Green hydrogen projects currently under way in the EU are just sufficient to meet the EU goal of having a 6GW installed capacity by 2024 and will greatly exceed the 40GW goal for 2030 with 139GW in projects planned to be realized in that timeframe.[34] This would be sufficient to supply the 10Mt of hydrogen needed for Ff55 and RePower EU. Around 56% of the projects are planned for Spain and Portugal, which have the best conditions for hydrogen production in the EU. In addition, as mentioned previously, the EU is building hydrogen production partnerships with Northern Africa, the Middle East, Australia, Chile and other countries.[35]

---

[31] In addition to the EU Hydrogen Roadmap see also *Energy systems integration – Hydrogen, European Commission.*

[32] Hydrogen Europe (2022). *Clean Hydrogen Monitor.*

[33] The levelized cost of hydrogen (LCOH) accounts for all the capital and operating costs of producing hydrogen, thus enabling different production routes to be compared on a similar basis.

[34] Hydrogen Europe (2022). *Clean Hydrogen Monitor.*

[35] For an analysis of the strategic value of the potential partners see Toward a hydrogen import strategy for Germany and the EU—Priorities, countries, and multilateral frameworks, Dawud Ansari and Jacopo Maria Pepe (2023), SWP Working Paper.

Table 2.3 Projections for hydrogen demand by sector by 2050

Sector	European Hydrogen Roadmap	Energy Transition Outlook (DNV)
Transport[a]	30%	40%
Industry	39%	43%
Buildings	26%	17%
Power generation	5%	–
Total energy demand	2251 TWh = 8.1 EJ	1250 TWh = 4.5 EJ

Sources: EU Hydrogen Roadmap, DNV, Allianz Research
[a]For more information on the transportation sector, see Chap. 3 Transportation

Investment in renewables must occur simultaneously with de-investment in fossil fuels for the energy transition to be successful and complete. As discussed previously, public and private investments will both play an important role, but governments should also reassess their fossil fuel subsidies as soon as the current energy crisis eases.

Slashing fossil fuel subsidies would not only reduce their impact on government budgets, but also give a significant boost to the green transition. The EU is taking steps to gradually get rid of energy subsidies (particularly fossil fuel subsidies) in line with commitments made in the Paris Agreement, G7 and G20 and the Green Deal's principle to "do no significant harm". Despite such ambitions, a 2022 report from the Commission showed an increase in the overall amount of energy subsidies (particularly fossil fuel subsidies) across the EU, except for a handful of member states: Austria, Denmark, Estonia and Hungary.

The EU's total energy subsidies reached €184bn in 2021, a 15.7% increase from 2015.[36] Demand subsidies that incentivize energy consumption (such as tax breaks or income support) grew by 14% from 2021. The types of subsidies and their allocation structures vary across the EU. For instance, Latvia mostly subsidizes energy efficiency measures, while Germany focuses its efforts on subsidizing renewables. At the other end of the spectrum, Belgium, Greece, Ireland, and Bulgaria spent the most on fossil fuels.

---

[36] European Commission (2022). 2022 Report on Energy Subsidies in the EU.

Conclusion

As ambitious as the goals of the EU may appear at first sight, they still fall short of what is needed to reduce its energy-related carbon footprint to a 1.5 °C-compliant pathway. All told, investments will need to be 84% higher than what is currently envisioned over the rest of this decade—just to close the 4-year implementation gap with what is needed to stay below the 1.5 °C mark.

If anything positive can be said about the Russian invasion of Ukraine, it would be that it disabused the EU of its notion that relying on cheap energy from an autocratic government was a good thing, while providing a massive boost to the deployment of renewable sources of energy.

Still, fossil fuels will continue to play an important role for many years yet. This calls for policy to strike a fine balance between removing fossil-fuel subsidies and minimizing hardship for both households and industry, while simultaneously introducing the right mix of incentives and disincentives.

At the same time, investors will need to strike a different, but equally fine, balance between de-investing in fossil fuels and investing in renewables. Among the latter, green hydrogen is expected to play a key role thanks to its versatility and potentially unlimited inputs: water and carbon-free electricity.

Getting the transition of the energy sector right is crucial for efforts to contain global warming. It is by no means enough on its own to check the temperature rise, but, given that it underlies all other sectors, it is the most pivotal. In particular, a successful transition to greener sources will help the next-biggest emitter accomplish its own transition: utilities.

3

The EU Utility Transition
Electrifying Times

Electricity is a very particular good. It is both a basic part of nature—the stuff of lightning as well as the medium by which neurons pass signals to the muscles—and the backbone of modern industrial society. It is also a most homogenous good, because there is no quality difference, and at the same time most heterogeneous thanks to its extraordinary versatility. Since storing it at reasonable cost is hard, supply must match demand at every point in time—and be delivered at the exact moment it is needed.

And despite its homogeneity, it can be generated from many different primary sources of energy. The trouble is that a large proportion of such primary sources are fossil fuels. Generating electricity from them emits greenhouse gases that, as we all know too well by now, warm up the atmosphere.

Still, electricity is *the* key to a less-hot planet. In fact, decarbonizing our economies hinges to the greatest extent on how far, and how fast, we can electrify our transportation, industry, buildings, and services. The corollary is that if we want to use electricity to do things currently done with fossil fuels, we will have to generate an order of magnitude more of it—without spewing more carbon into the atmosphere.

Thus, clearly, the power-generation sector's transition to renewable energy sources will play a critical role in emissions reduction. Even though final energy demand is expected to decrease over time, global primary energy production will still need to increase, since using electricity as storage and to produce synthetic fuels faces conversion losses. In 2015, electricity was responsible for around 23% of final energy demand in the EU, a share that is

L. Subran, M. Zimmer, *Investing in a Changing Climate*, Professional Practice in Governance and Public Organizations, https://doi.org/10.1007/978-3-031-47172-8_3

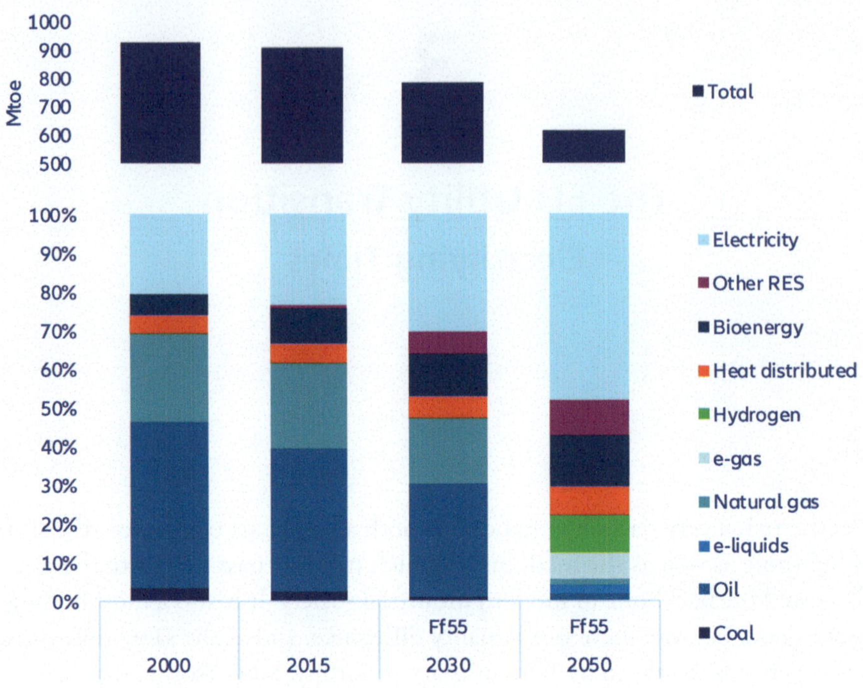

Fig. 3.1 Final energy demand in the EU by energy carrier. Sources: Allianz Research, European Commission 2030 Climate Target Plan

expected to rise to 29–31% in 2030, and to 46–50% by 2050—with an additional 20% coming indirectly from using it to produce fuels, such as hydrogen, e-gas, and e-liquids (see Fig. 3.1 below).

Electrification is being driven by demand growth primarily from the transportation sector, followed by the industry and residential sectors (Fig. 3.2).

The EU's recently proposed 'Fit for 55' (Ff55) legislation[1] expects demand from the transportation sector alone to increase by a factor of 2.5–2.9, with the growing number of electric vehicles and the roll-out of charging infrastructure[2] expected to require an additional 100 TWh by 2030 and 484 TWh by 2050, compared to 2019 levels.[3] Other sources even suggest that an additional 760 TWh will be needed.[4] The residential sector will see the emergence of

---

[1] The Fit for 55 legislation, announced in July 2021, aims to reduce greenhouse gas emissions by 55% by 2030.

[2] For more on transportation, see Chap. 3 Transportation.

[3] For comparison, the increase from 2015 to 2019 was about 3.6TWh.

[4] ENTSO-E & ENTSOG TYNDP. *Visualisation Platform.*

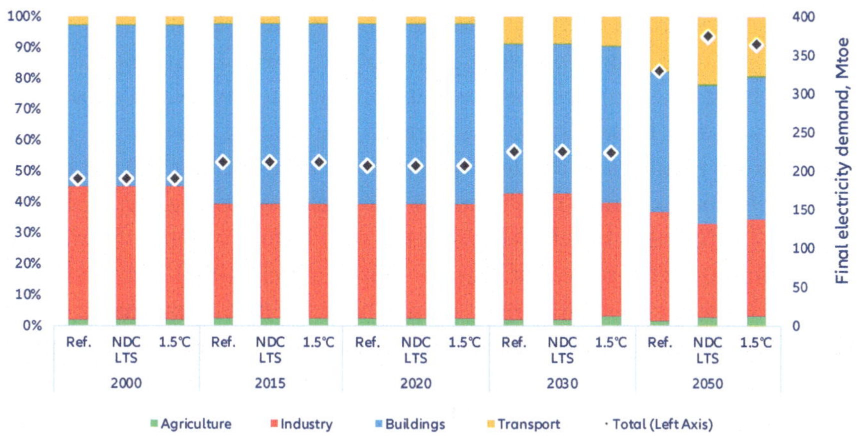

Fig. 3.2 Final electricity demand (EU-27) by sector. Sources: European Commission JRC GECO

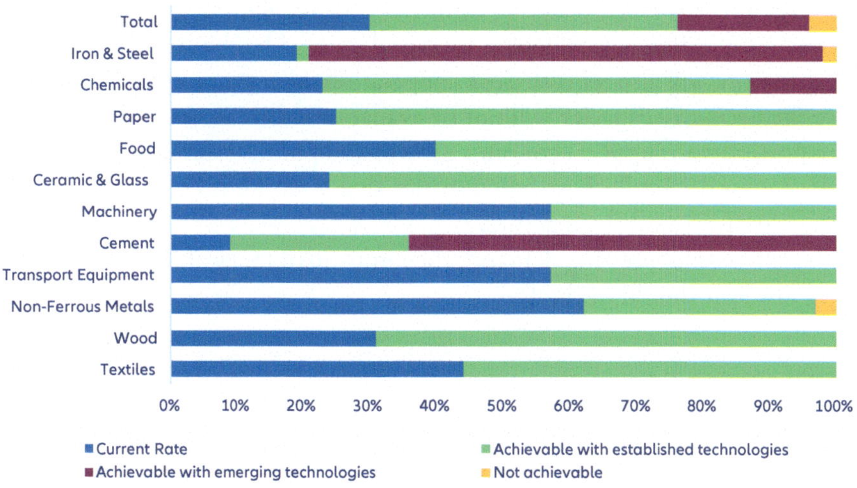

Fig. 3.3 Potential electrification in EU industries. Source: ETIP Wind

electric heat pumps for heating and cooling as the main driver, with the share of electricity in residential energy demand likely to rise from 25% today to 45–60% by 2050. The industrial sector's electricity consumption could rise to 76% of its total final energy consumption if established technologies are incorporated into current processes, which would contribute to additional reductions in greenhouse gas (GHG) emissions (43%) and energy losses (7%)[5] (Fig. 3.3).

---

[5] ETIP Wind (2021). *Getting Fit for 55 and set for 2050.*

But how exactly can this rising demand for electricity be met?

The answer is through wind and photovoltaic power. Technological maturity and rising CO_2 prices—not to mention the current spike in fossil fuel prices, which will hopefully be temporary—are driving down their relative costs, making them ready for deployment across Europe.[6] At the same time, the electrification of industries requires the deployment of green hydrogen, produced from renewable electricity. In the steel industry alone, for example, transitioning to net-zero would require approximately 400 TWh of electricity, which is seven times the amount it uses today. Of this, an estimated 62.7% (~250 TWh) would be used to produce 5.5 m tons of hydrogen. Green hydrogen will also be crucial for decarbonizing the chemical industry, especially in ammonia production. Considering that demand will grow as ammonia is utilized as an alternative fuel for shipping, the production of ammonia from green hydrogen is critical.

However, ramping up green hydrogen is a complex task, demanding tight coordination between European countries and regulations in which carbon contracts for difference will play a central role.[7] Developing the necessary infrastructure will also require close cooperation with regions outside the EU, since the necessary capacity will not be available within the EU. This provides an opportunity to support economic growth and political stability, for instance in African and Middle Eastern countries that have ideal conditions for the expansion of renewable capacities.

Overall, despite pressing demand, the ramp-up of renewables is not unfolding as fast as it should in the EU.

---

[6] It is often questioned whether there is enough room and enough wind and sunshine in Europe to make this possible. The short answer is: Yes. See for instance FFE (2020). *Info: Electrification vs. vRES potentials in Europe – Is the potential for variable renewable energy sources sufficient to cover post electrification electricity demand?*.

[7] 'Carbon contracts for difference' (CCfD) exist in various national implementations. The German government, for instance, announced a CCfD concept in which energy-intensive industries will be compensated by climate protection agreements for a period of 15 years to cover for their additional costs (OPEX and CAPEX) to convert their production to climate neutrality. These climate protection agreements thus make green technologies more attractive for energy-intensive industries. CCfDs payments are conditional on certain market prices—for instance the hydrogen market price—and cover the price difference to an agreed 'strike price'.

The Rise of Renewables and Fall of Fossil Fuels

For renewable electricity, 2020 was a landmark year. For the first time, renewables overtook fossil fuels to become the main source of electricity in the EU (38%).[8] Together, wind and solar generated one-fifth of the EU's electricity, while coal contributed 13% (nearly halving since 2015). In terms of total generation, wind led the way (+5% since 2015, to 14% in 2020), followed by solar (+2% since 2015, to 5% in 2020).

Conversely, hydropower and bioenergy have stalled, and 2020 also saw the second-largest decrease in nuclear generation since 1990, with a 10% reduction that was only exceeded by the 16% drop occurring in 2022.[9] This highlights the need to ramp-up solar and wind even further.

To fulfil the EU's Ff55 ambition of reducing greenhouse gas emissions by 55% by 2030, renewables will necessarily have to play a large role, helping to reduce GHG emissions by around 70% (compared to 2015 levels) in the power-generation sector. By 2030, the electricity sector will see the highest share of renewables, with over 60% in all Ff55-compliant scenarios. By 2050, power generated from renewables is expected to exceed 85% (Fig. 3.4).

Electricity can be measured in two different ways: by installed capacity, and by production. Installed capacity, normally measured in gigawatts (GW), refers to the maximum output of electricity that can be produced under ideal conditions. Production, in turn, refers to the amount of electricity produced over time in terawatt hours (TWh). The share of solar and wind production is expected to grow the most, from 13% in 2015 to 48% in 2030, and eventually to 67% by 2050 (Fig. 3.5).

Wind energy will be the largest source, providing around 34–35% of all electricity in 2030, which calls for it to have the highest installed capacity over the long term, mostly onshore: more than 1000 GW by 2050 compared to 2015 levels (Fig. 3.6).[10]

---

[8] AGORA (2021). *The European Power Sector in 2020.*

[9] These plans might come under review in the context of the current energy-price crisis. In October 2021, for example, French President Emmanuel Macron announced plans to expand nuclear power and invest in small modular nuclear reactors (EUR1 billion until 2030), France 2030 Plan. For additional data see EMBER. *Nuclear Overview.*

[10] Recent studies indicate wind onshore plus offshore will need to increase to over 1200GW in 2050 in total, compared to 145GW in 2015. See ENTSO-E & ENTSOG (2022). *TYNDP 2022* and FFE. *eXtremOS Project.*

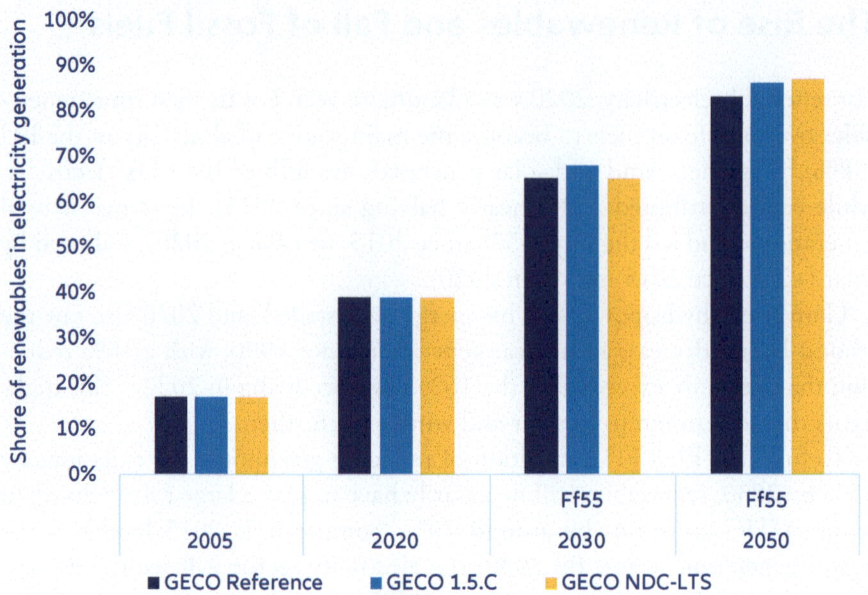

Fig. 3.4 Share of renewables in electricity generation (EU-27). Sources: Allianz Research, European Commission JRC GECO

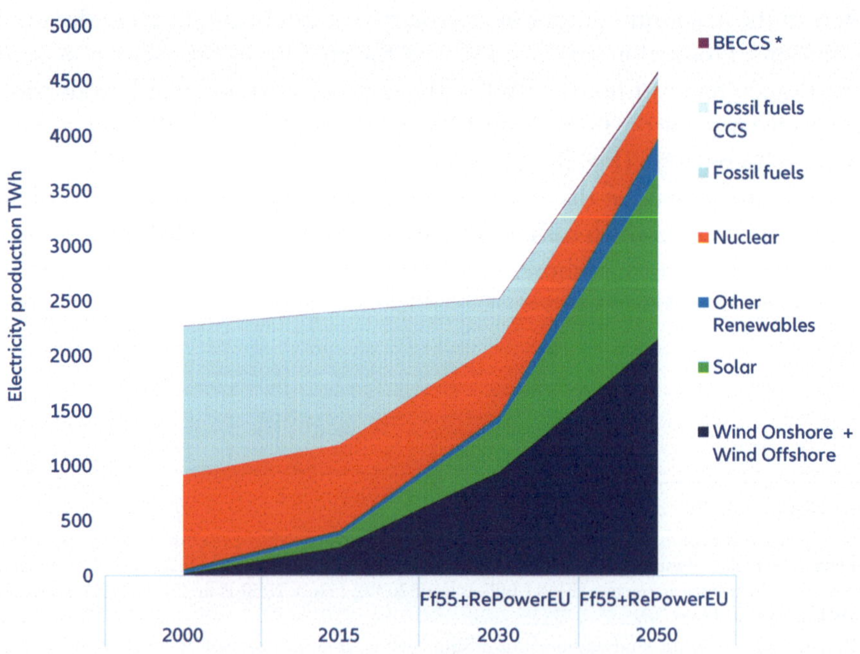

Fig. 3.5 Electricity production in EU-27, TWh. Sources: Allianz Research, European Commission JRC GECO, * BECCS is bioenergy combined with carbon capture and storage

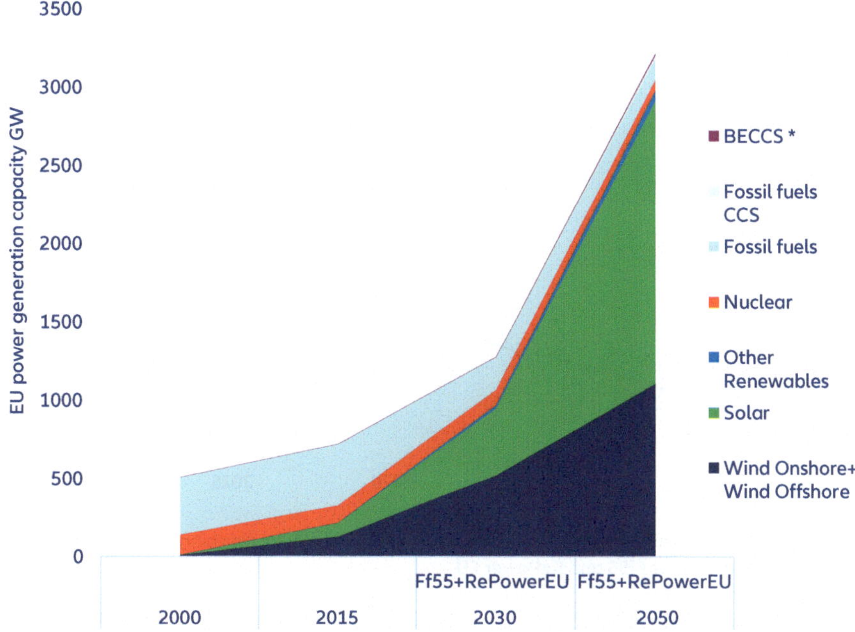

Fig. 3.6 Installed power generation capacity in EU-27, GW. Sources: Allianz Research, European Commission JRC GECO, * BECCS is bioenergy combined with carbon capture and storage

From 2010 to 2018, renewable electricity growth clocked only around 3% per year (approximately 38 TWh/year), which is not enough to meet future needs. Indeed, growth must increase threefold in order for the 60% share to be met by 2030. In addition, the Ff55 and RePower EU ambitions will still have to be carried over to the EU member countries' national energy climate plans (Fig. 3.7).

However, there is often an overlooked trade-off that comes with expanding wind and solar PV: increased competition for space and land, which adds to biodiversity concerns. Land is valuable and the agriculture industry is a critical component of our ecosystem—its role in storing carbon, providing a habitat for biodiversity and as the foundation of our food system is invaluable, but this does not mean it cannot be used in sector-coupling with energy production. This will require a rethinking of agricultural zoning with respect to the co-use of land for agriculture, energy production and carbon storage.[11] Goals for co-use include agriphotovoltaics (the combined use of agricultural land for

---

[11] See Chap. 6 Agriculture for more information on nature-based carbon capture and storage.

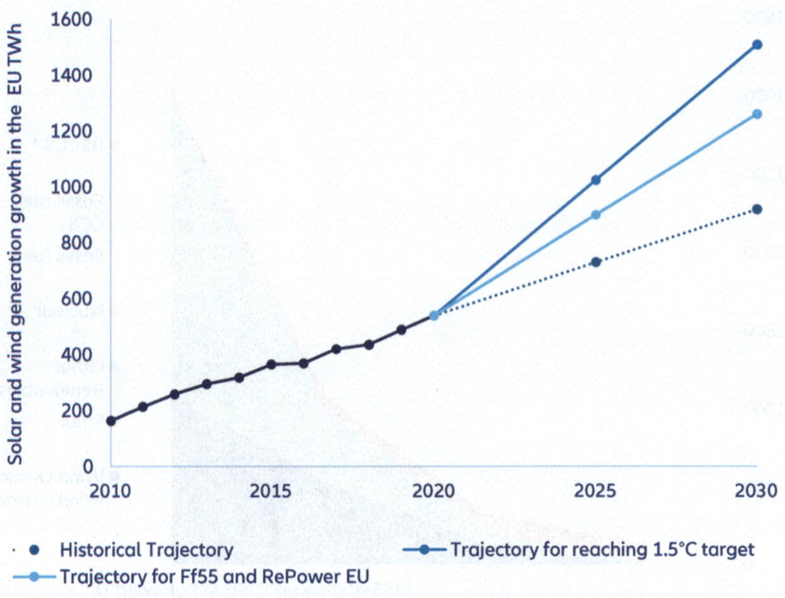

Fig. 3.7 Solar and wind generation growth in EU-27. Source: Allianz Research

photovoltaic energy production, e.g., by installing solar panels on stilts above meadows) as well as agroforestry (e.g., planting energy tree strips in form of short rotation plantations on cropland). The potential solutions to these conflicts, such as rooftop-photovoltaics or the above-mentioned agriphotovoltaics, come with additional costs. Therefore, an essential enabler will be adequate support mechanisms and regulatory adjustments to provide sound revenues to incentivize investments in co-use activities.[12]

Looking at individual member countries, we find clear leaders and laggards when it comes to solar and wind development. For electricity production, Denmark is a clear EU leader, as it generated 62% of its electricity from wind and solar in 2020. Following (rather far) behind are Ireland (35%), Germany (33%) and Spain (29%)[13] (Fig. 3.8).

In contrast, Italy, Bulgaria and the Czech Republic have seen very limited growth since 2015, despite having excellent conditions for solar and wind power

---

[12] These regulations and subsidy programs are enacted at the national level, and issues can differ in each EU country. Typically, a co-use would mean the loss of agricultural subsidies or require a different zoning for the activity (which makes it regulatory impossible to have both uses on the same piece of land). Subsidies might also depend on the type of agricultural co-use, thus paying higher renewable subsidies for farming than for livestock and therefore imposing additional implicit restrictions on the economic utilization of the land.

[13] AGORA (2021). *The European Power Sector in 2020.*

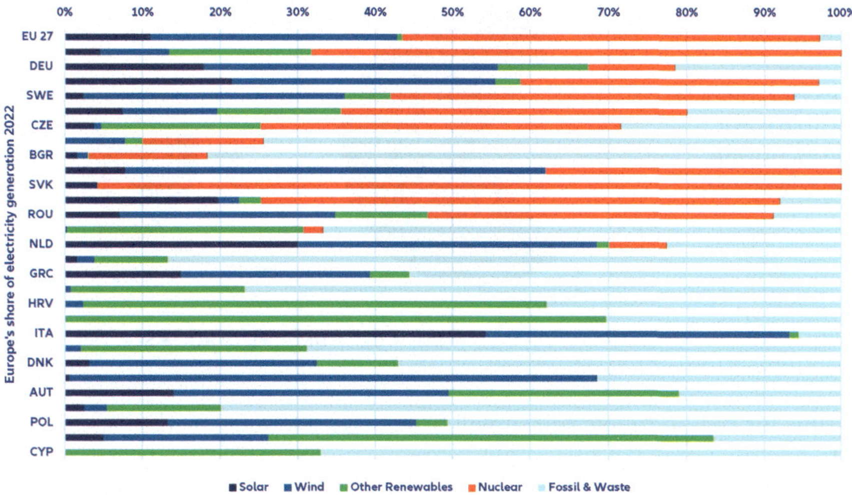

Fig. 3.8 Europe's share of electricity generation in 2022, by country. Sources: EMBER, Allianz Research

generation. But even "leaders" still have a long way to go, as the phasing-out of coal remains a key challenge in some of them. In all main policy scenarios, coal will need to be completely phased out by 2030 to achieve a 55% reduction in emissions. Although coal-fired power generation fell by 20% on average across the EU in 2020, laggards such as Poland only achieved an 8% reduction. And most coal-reliant countries are not planning a complete phase-out before 2038: While Germany has announced a more ambitious 65% national renewable target for electricity by 2030, it is also one of the six EU member states that has not yet committed to phasing out coal before 2030, along with Poland, Romania, Slovenia, Bulgaria and the Czech Republic. Based on current national policies, 38 GW of coal capacity will remain after 2030 in these countries.

The reliance on coal persists because there is a tradeoff between GHG emissions-reduction and stabilized supply in Europe: Generating power from wind and solar PV is highly dependent on weather conditions, with fluctuations inevitably occurring throughout the year. *Dunkelflaute*, the German term to describe the time of year when there is little wind and sun to generate power, is a real concern for the European continent in the winter months, especially since that is the time of year when energy demand also peaks. This, combined with below-par weather conditions, could exacerbate the intermittency of renewable electricity, which in turn could result in volatile energy prices. In the case of Germany, the balancing power to ensure supply security during *Dunkelflaute* is provided now by lignite-, hard coal- and gas-fired power plants. Nuclear power has already been phased out, and to make it

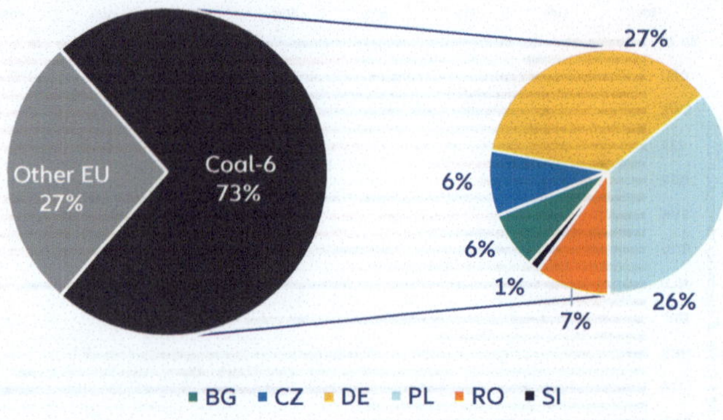

Fig. 3.9 Distribution of additional investment (EUR131bn total) for a 2030 coal exit. Sources: AGORA, Allianz Research

possible for coal to follow suit, it will be necessary to increase the capacities of hydrogen-ready gas-fired power plants.[14]

As a result, back-up renewable capacities are needed: after all, baseload power plants like coal-fired power stations with more than 5000 yearly full-load hours (FLHs) will get decommissioned (because renewable production reduces the demand for their FLHs, which makes them unprofitable) or repurposed as subsidized capacity reserve with much lower FLHs for wind power plants with typically less than 2000 FLHs (though offshore wind facilities clock up to 4000 FLHs), or photovoltaic installations with typically 800–1200 full-load hours.[15] In this context, total installed capacity needs to increase by more than twice the rate of generation. To phase out coal by 2030, an additional 100GW of wind and solar PV capacity is needed across the EU, as well as 15GW of flexible gas-fired power plant capacity.[16] This requires approximately EUR131bn in additional cumulative investment across the EU until 2030, with 63% (EUR83bn) going towards wind, 23% (EUR30bn) towards solar and 14% (EUR19bn) towards new gas plants. Of this, 73% (EUR96bn) is required by the six most coal-reliant countries, with Germany (EUR35bn) and Poland (EUR34bn) needing the most additional investment to complete their coal exit (Fig. 3.9).

---

[14] However, they are not nearly as flexible as gas power plants (see Appendix on *load gradients, minimum load and start-up times of power plants of different technologies*). Gas-fired power plants are the better option for energy systems with high penetration of intermittent renewables.

[15] For more information, see Mats de Groot, Wina Crijns-Graus, & Robert Harmsen (2017). *The effects of variable renewable electricity on energy efficiency and full load hours of fossil-fired power plants in the European Union.* Energy, 138, 575–589. https://doi.org/10.1016/j.energy.2017.07.085, EnergyBrainpool (2017). *Flexibility needs and options for Europe's future electricity system* and Matthias Huber, Desislava Dimkova, & Thomas Hamacher (2014). *Integration of wind and solar power in Europe: Assessment of flexibility requirements.* Energy, 69, 236–246. https://doi.org/10.1016/j.energy.2014.02.109

[16] AGORA (2021). *Phasing out coal in the EU's power system by 2030.*

Carbon Pricing for a Market-Based Coal Exit

Carbon pricing will also play a key role in driving renewables supply and incentivizing a coal exit by 2030. AGORA's analysis for a complete 2030 coal exit identifies three possible policy scenarios, with carbon price projections by 2030 (Fig. 3.10).

In the first, the "market-based coal-to-clean" (MCTC) scenario, both coal exit and expansion of renewables are driven by the EU Emissions Trading System price (ETS). In the second, the "market-based coal exit" (MCE) scenario, only the coal exit is driven by the EU ETS, while renewables expansion is driven by additional EU or national subsidies. In the third, the "policy mix" (PM) scenario, EU ETS prices are insufficient to ensure a market-driven coal exit by 2030, making it necessary for national mandatory decommissioning policies to complement the efforts and for additional subsidies to be offered for a sufficient expansion of renewables. The scenarios suggest that for a market-based coal exit as in Scenario 2, a carbon price floor of EUR65 per ton of CO_2 is needed by 2030, a value that was in fact already exceeded in November 2021; the carbon price per ton has been above EUR80 since the beginning of 2023.

Based on this, if Scenario 1 is followed and the carbon price reaches approximately EUR152 per ton in 2030, the price of lignite would be the most affected. The base price of EUR50 per ton SKE (from the German *SteinKohleEinheit*, meaning hard-coal unit, which expresses an energy content equivalent of burning 1 kg of hard coal) would see the current carbon price premium (assessed at EUR80/tCO_2) almost double, to EUR500 per ton SKE.[17] On the other hand, such a carbon price would cause the carbon price premium for gas to also almost double, to about EUR30 per MWh of primary energy content. This has to be put into perspective against the gas price, which jumped from EUR16 per MWh in February 2021 to EUR340 per MWh in August 2022, and down again to EUR26 in July 2023.[18]

---

[17] Source for base price assumption: AGORA and Umweltbundesamt. Hard coal is more expensive, with local German production costs of above EUR180/ton or EUR180/tonSKE.

[18] Assuming a current EU ETS price of EUR60 and emissions per primary energy content of around 201 kg/MWh for gas and of 3250kgCO2/tonSKE for lignite. This is per kWh of electricity produced, equivalent to around 433kg/MWH for gas and 1093kg/MWh for lignite.

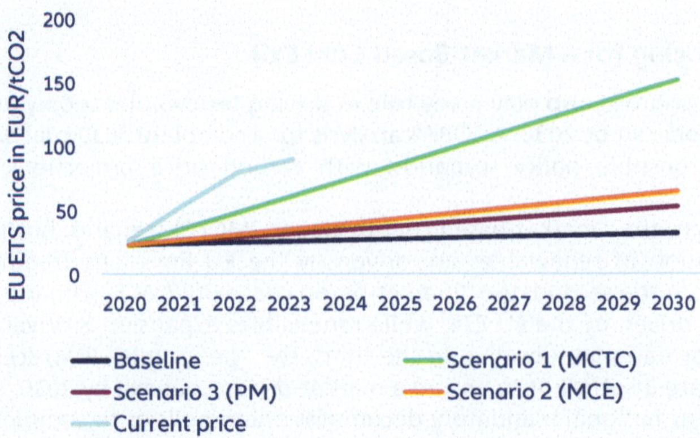

Fig. 3.10 CO_2 price projections for a coal exit by 2030, by scenario. Sources AGORA, Allianz Research

New gas-fired power plants are needed primarily to cover the system's flexibility demand: in times of low renewable electricity availability, gas-fired plants step in to fill demand, replacing coal to stabilize electricity supply.[19]

These new gas power plants should also be "hydrogen-ready", i.e., designed to be converted and retrofitted to run on (green) hydrogen once a market is developed for its production and distribution. Still, it should be noted that this is a mid-term solution with limited utilization: From 2023 to 2030, the utilization of these gas-fired power plants is expected to decrease from around 3000–3400 to 2000 full-load hours yearly, which means that by 2030, these plants will only be utilized around 22% of the time (Fig. 3.11).

The decrease in full-load hours results in an increase in the levelized cost of electricity (LCOE)[20] of around EUR20/MWh, to approximately EUR100/MWh (Fig. 3.12).

Yet the bill for the coal exit doesn't stop there. In order to be on the safe side, strategic reserves are needed. At present, policies for strategic gas reserves are up to the individual member states to decide: there is no EU coordinated approach. AGORA estimates that if countries want to cover domestic peak loads with their own capacities, they will require additional adjustments to the electricity system and an extra 4 GW of strategic reserves: 3.4 GW for

---

[19] See Appendix on *load gradients, minimum load and start up times of power plants of different technologies* to see how well-suited gas power plants are for volatile electricity demands.

[20] A measurement of total cost divided by energy/electricity generated by an asset over its lifetime.

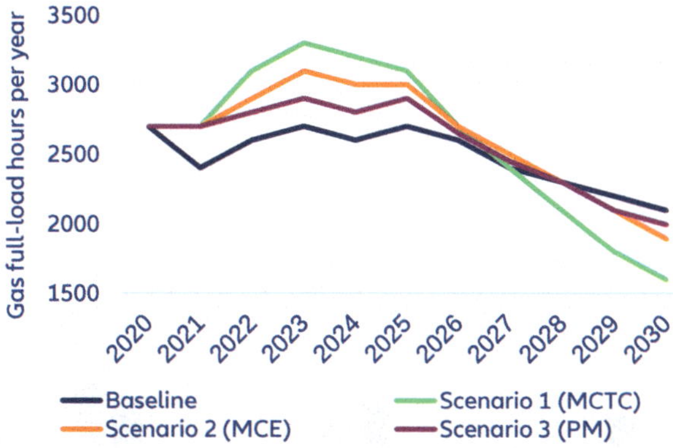

Fig. 3.11 Utilization of gas power plants at full-load. Sources: AGORA, Allianz Research

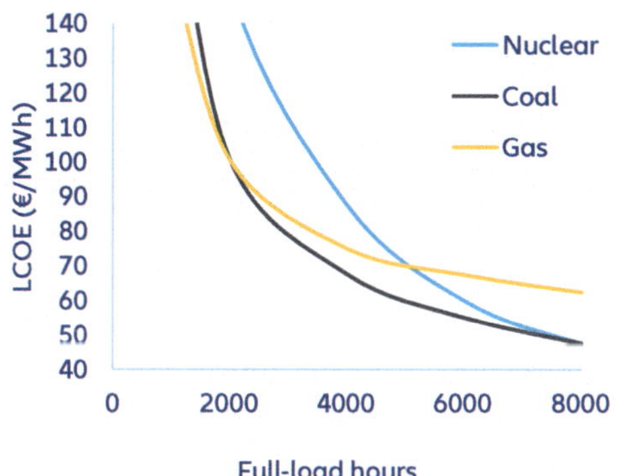

Fig. 3.12 Levelized generation costs by technology. Source: Allianz Research

Germany,[21] 600 MW for Bulgaria, 100 MW for the Czech Republic, and 300 MW for Romania. This would call for an additional EUR14bn, bringing the total to an additional EUR145bn needed until 2030.

Storage solutions can also provide additional stability and flexibility to the electricity supply chain. Currently, the most common storage solutions are pumped hydropower and batteries, but the fluctuating nature of wind and solar power will lead to fluctuating electricity prices, thus incentivizing the flexibilization of electricity demand from industry and transport. These

---

[21] For reference, Germany already has approximately 30GW of net installed electricity generation capacity for gas, which represented about 13% of total installed capacity in 2021.

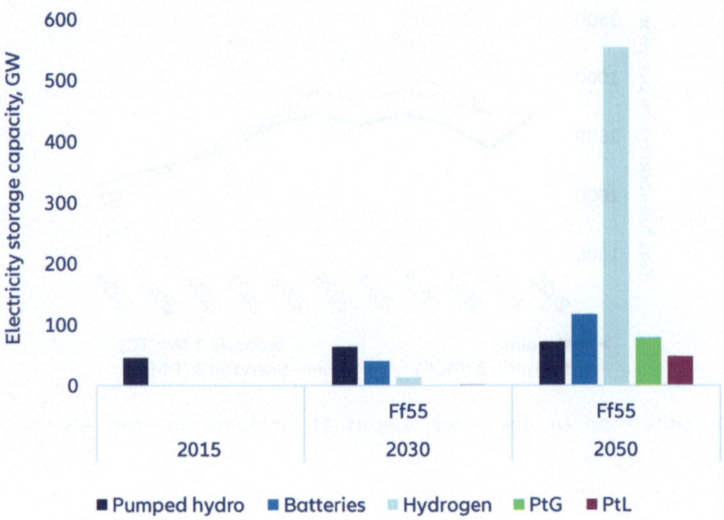

Fig. 3.13 Electricity storage capacity. Source: European Commission 2030 Climate Target Plan and JRC GECO

demand-side responses will be aided by storage technologies like batteries, or by converting electricity into hydrogen, a key fuel that can help to balance fluctuations in production caused by meteorological intermittencies. The hydrogen thus produced can subsequently be converted via power-to-gas (PtG) or power-to-liquid (PtL) processes for easier transportation and storage. As a result, installed electrolyzer capacity to produce hydrogen is expected to grow from around 12–13 GW to 528–581 GW from 2030 to 2050 (Fig. 3.13).

The expected expansion of hydrogen production and storage capacity complements well the "hydrogen-ready" gas power plants that will be used for power-supply stabilization. Ramping up hydrogen production, however, is particularly challenging not only financially, but also from the perspective of a coordinated expansion of supply, demand, and transportation infrastructure. While carbon prices are the most powerful and effective way to advance the green transition, there are some cases—for example path dependencies resulting from long investment cycles or the establishment of markets for new technologies—where additional instruments might be required, such as the so-called carbon contracts for difference (CCfDs). When infrastructure projects require a carbon price beyond the prevalent levels to be competitive versus their fossil fuel-based alternatives, carbon contracts for difference provide an additional financial benefit that compensates for the difference and makes the project attractive for investors. CCfDs are thus a very useful—if not crucial—instrument for establishing a viable hydrogen market.

In the end, market flexibility is the best solution for intermittent solar- and wind-based power generation. Although Europe's internal electricity market has been liberalized since the late 1990s, countries are also taking additional steps to support the EU market, as their national markets have benefited from using cross-border electricity trade and from lower prices. The merging and integration of regional and national markets, as well as their digitalization, needs to be strengthened further, since this could increase security of supply and decrease costs via trading. Significant investment is still needed to make grids smarter, increase transmission capacity and strengthen network infrastructure to ensure that countries can take advantage of future electricity trading (investment figures are explored further in later sections of this chapter).

The Investment Gap: Fit for 55, But Not for 1.5°

In the EU, planned investments into the energy system still fall short of what is needed. In the power-generation sector, IRENA estimates that the approximately EUR84.7bn per year already planned to ramp up renewables and address power grids and system flexibility will not be enough to support a net-zero economy. It calculates that an additional average of around EUR40.7bn per year is needed until 2050,[22] for a total investment of EUR125.4bn per year. Support for renewables should be about EUR67.4bn per year, while EUR48.4bn per year should go towards power grid flexibility. The Ff55 proposal, for its part, estimates that even larger investments are needed, amounting to some EUR58.8bn per year until 2030, together with EUR81.7bn per year until 2050 for power grids, which is in line with estimations by ETIP Wind.[23]

One of the biggest challenges for the power grid is capacity expansion and optimization to allow for smooth cross-border electricity flows. Today, Europe has approximately 50 GW of cross-border capacity, which is not enough: the European Network of Transmission System Operators for Electricity (ENTSO-E) has called for 85 GW of *additional* capacity by 2030. Current projects in the pipeline should add an additional 70 GW by 2030, worth approximately EUR50bn, which is still 15GW short of the ENTSO-E's proposal. In Germany alone, an estimated investment of over EUR100bn is

---

[22] IRENA (2020). *Global Renewables Outlook: Energy Transformation 2050.*
[23] Estimated that grid investments need to average between EUR66–80bn per year over the next 30 years (ETIP Wind (2021). *Getting fit for 55 and set for 2050.*

needed to optimize and develop the existing AC transmission grid, add devices to steer power flow, and add new DC connections to support domestic and cross-border electricity transmission needs.[24] Achieving these depends on progress in digitalization and the availability of skilled personnel. The roll-out of smart meters and the increasing numbers of prosumers—energy customers who both consume and produce—or community electricity models require adequate regulatory frameworks for data protection, building law and zoning, a reduction of bureaucracy and the acceleration of approval processes.

As Fig. 3.14 shows, electricity production will not only need to shift to renewable energy sources, but must also increase across all European countries to achieve the 1.5 °C goal. Consequently, power plants will need investments to the tune of EUR57bn per year until 2030, followed by EUR89.4bn per year until 2050. In addition, around EUR602.5 m per year is recommended to ramp up electrolyzers for renewable hydrogen production.

As previously mentioned, carbon pricing is the most effective tool to speed up the energy transition. Another instrument is a credit-enhancement arrangement between a public player, such as a development bank, and a private investor. Public institutions must explore and use these public-private partnerships to the greatest extent possible, because crowding in private capital is key to the green transition, especially as regards enhancing speed: the secret to opening up markets to new technologies is a quick scaling-up to drive costs down. For that, a combination of the risk capacity of the public sector and the knowhow and capital of the private sector is essential.

Generation costs of renewables are continuing to drop as a direct result of the ramp-up of renewable capacities. ETIP estimates that by 2050, all wind energy forms will have a levelized cost of electricity (LCOE) below EUR53/MWh, with onshore wind LCOE expected to be EUR33/MWh by 2030. In Germany, by that year, the LCOE for onshore wind is expected to range between EUR30 to 70/MWh, while PV may range from EUR20 to 80/MWh.[25] This is in line with the stable relationship between total global

---

[24] Grid Development Plan 2030, Netzentwicklungs Plan Strom (2023), Page 253.

[25] Frauenhofer ISE (2021). *Levelized Cost of Electricity- Renewable Energy Technologies.*

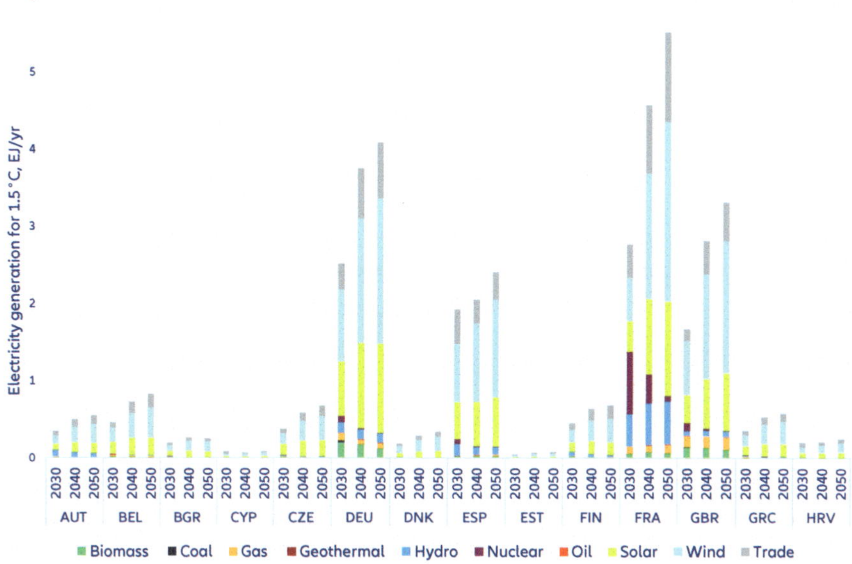

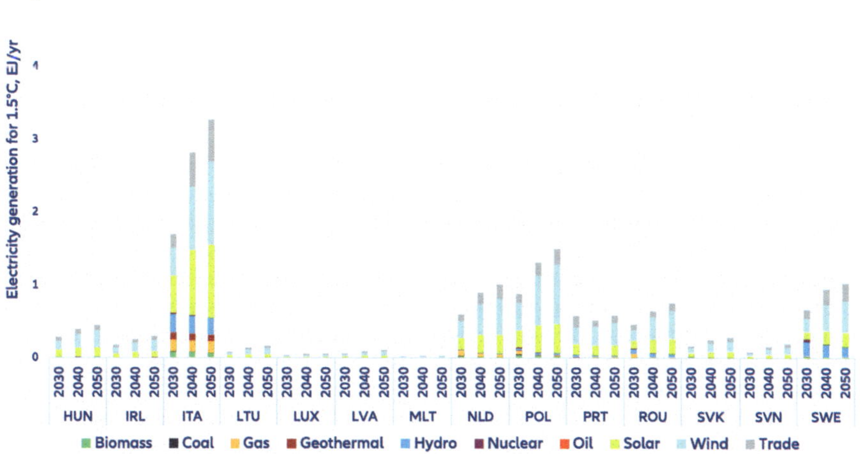

Fig. 3.14 Evolution of electricity generation by technology in the EU-27 and the UK for reaching the 1.5 °C goal. Source: Allianz Research

installed capacity and LCOE. The learning curve describes the cost reduction for a doubling of the installed capacity and lies between 10% and 36% for the technologies considered (Fig. 3.15).

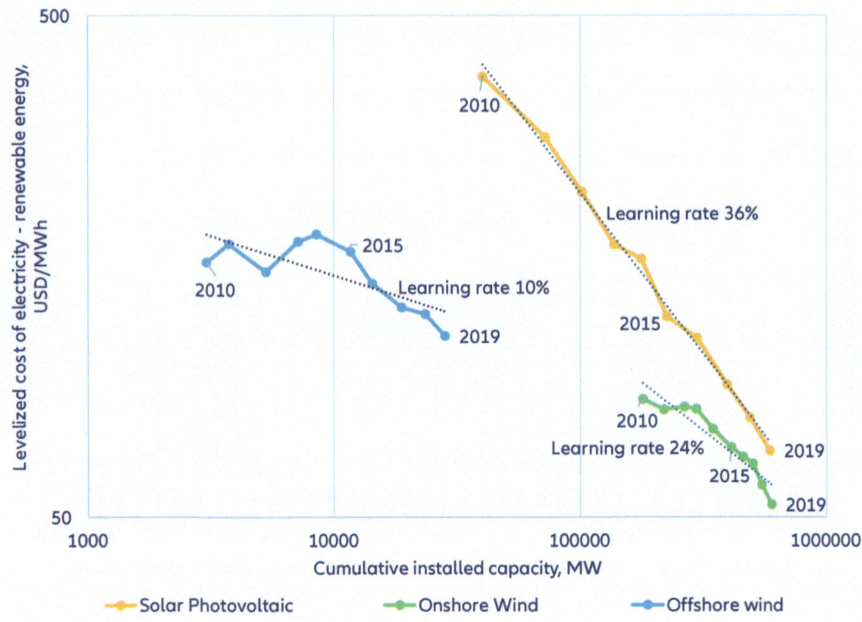

Fig. 3.15 LCOE of renewables (logarithmic scale). Source: Our-Word-in-Data, Allianz Research

What Does the Energy Transition Mean for Electricity Prices for Consumers?

A common and valid concern in the energy transition discussion is the effect on electricity prices for consumers. Rising prices for fossil-fired heating and gasoline will help reduce energy consumption in households, putting a strain on low-income households. Considering for instance the average emissions of a German citizen, of around 9 tons of CO_2 per year, and a cost pass-through of EUR50 per ton of CO_2, this could add up to EUR1,800 per year to a four-person household that is unable to adapt to lower emissions. It is therefore advisable to use the revenues from carbon-pricing policies to offset financial hardship, securing a just transition and increasing acceptance.

This is the double dividend of carbon prices: mitigating harmful greenhouse gas emissions by increasing emission costs, while also generating revenues that can be used to financially compensate particularly vulnerable households. The latter should be done in two forms: direct lump-sum transfers, like the so-called *Klimaprämie* (climate bonus) that has been repeatedly discussed in Germany, and a stabilization of electricity prices, which would also particularly benefit lower-income households and, not least, also the industry, thus protecting jobs and income for households. Based on the impact assessment of the Ff55 proposal, price increases compared to 2015 for private consumers and small and medium enterprises are likely to persist until 2030 (Fig. 3.16), when prices might come back to normal compared to the 2022–2023 spike in electricity prices caused by the current drop in Russian energy flows.

By 2050, prices are expected to be approximately 35% higher than if current policies were to continue (the baseline 'BSL'). The rate at which industry electrifies its processes, which is key to decarbonization, is also heavily impacted by electricity prices, along with a countervailing portfolio of support measures, including subsidies, loans and preferential tax write-offs. Investments in electrification require electricity prices to increase much more slowly than the costs for emission-intensive alternatives. This includes the expectations for electricity price developments. Thus, a credible regulatory framework is needed that decouples average electricity price increases for consumers from carbon price increases.

Still, price signals on the electricity market need to keep signaling phases of insufficient supply in order to stimulate investment in demand-side flexibility. Electricity price increases could—for instance—be offset by decreasing taxes and levies on renewable electricity. While over the first half of 2021, taxes contributed to an average of approximately 39% of the total price of electricity for consumers in Europe, this share went down to 15% in the second half of 2022, mainly driven by governmental support schemes in reaction to the energy crisis; such measures led even to net negative taxes in some countries (Fig. 3.17). Germany experienced the highest cost, with household electricity prices peaking at 46 cents/kWh in the first half of 2023, with taxes and levies making up 27% and network charges 21% of the price.[26]

Ff55 also proposes a review of the EU's energy-taxation framework, which was last updated in 2003. One of the key measures proposed that will be included in the review is that fuels will be taxed according to their environmental performance and energy content, rather than their volume, including the phasing-out of exemptions for certain products. This should make renewable fuels much more price-competitive for consumers.

National actions are also underway: Germany recently abolished its renewable electricity surcharge (EEG), which was cut from 6.5 cents/kWh to 3.7 cents/kWh in the beginning of 2022 and then dropped completely in the second half of that year, thus providing some relief for private households and SMEs. As this tax has been instrumental in funding the expansion of renewables over the years, the government needed to pick up the tab through the general budget and income from the national emissions-trading scheme for transport and buildings. The ramp-up of renewables also has the potential to add an additional 400,000 to 700,000 jobs in Germany (under current policies and net-zero scenarios, respectively) by 2030, mostly in the bioenergy, wind and solar sectors.[27]

---

[26] Eurostat data and BDEW (association of the German energy suppliers) deviate significantly in reported household electricity prices. While Eurostat states a price of EUR33.57 for the second half of 2022, BDEW reports household electricity prices at EUR40.07 for the same period.

[27] Katharina Utermöhl and Markus Zimmer (2022). *Germany's Easter Package: Great green intentions.*

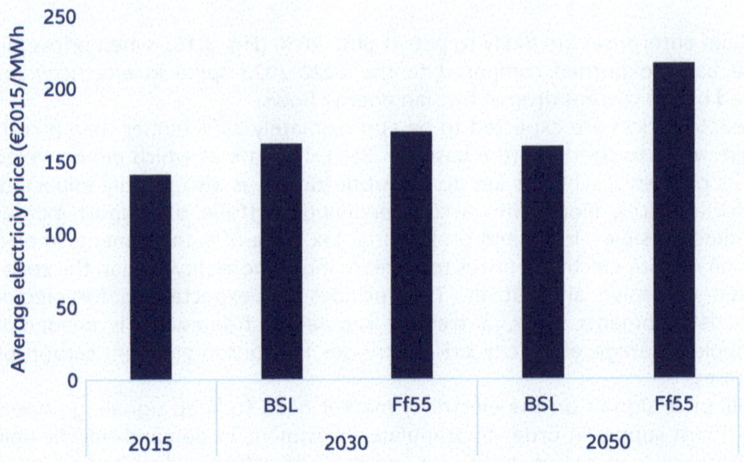

Fig. 3.16 Forecast for the average price of electricity in EU-27. Sources: Allianz Research, European Commission 2030 Climate Target Plan and JRC GECO

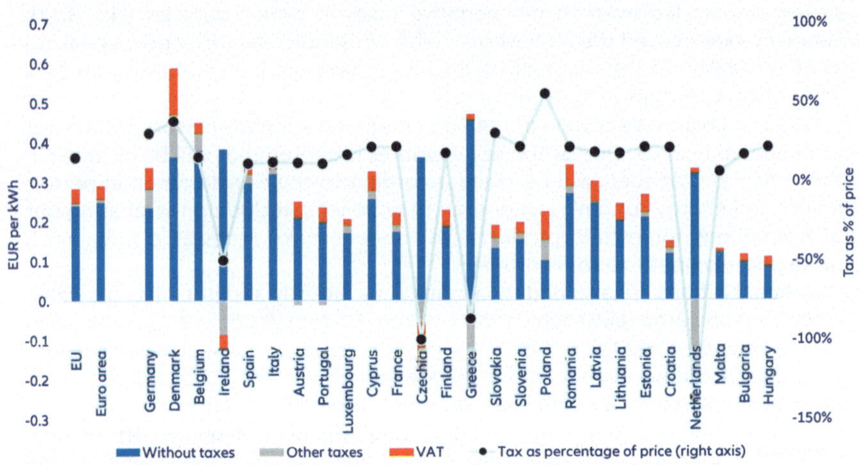

Fig. 3.17 Second half of 2022 breakdown of household electricity prices in the EU-27. Source: Eurostat

The Role of Emissions Trading and Carbon Capture & Storage

The emissions reduction proposed by the Ff55 regulation is ambitious, but additional revisions to the EU Emissions Trading Scheme (EU ETS) would help in strengthening an already established instrument for decarbonization.

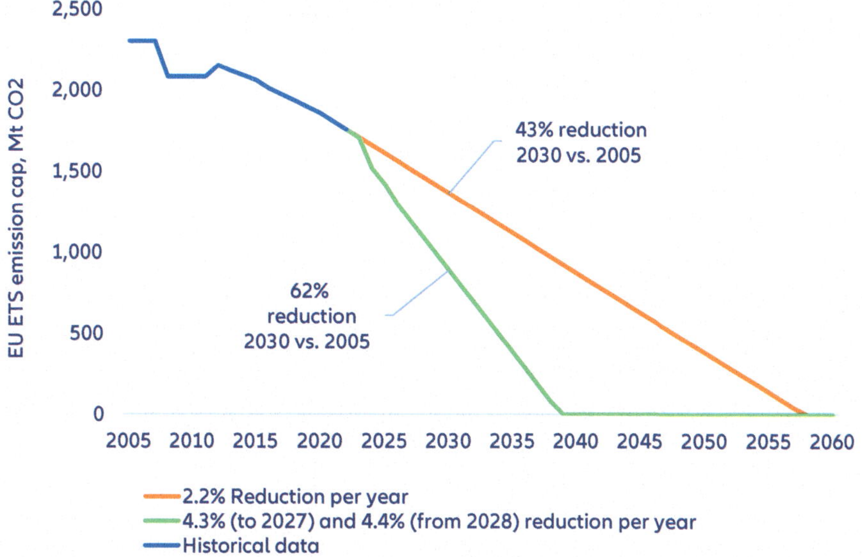

Fig. 3.18 EU ETS emissions reduction cap. Source: Allianz Research

The EU ETS follows a cap-and-trade approach, where a cap is placed on annual emissions across the EU, and companies must possess an allowance for each ton of CO_2 that they emit each year. Currently, the EU ETS covers around 41% of the EU's GHG emissions and applies to more than 11,000 power plants and industrial plants. With the new proposed legislation, the emissions reduction target for the EU ETS would change to 62% (previously 43% reduction) by 2030 compared to 2005 levels. This target means that the emissions cap must be lowered by 4.3% (previously a 2.2% linear reduction factor) each year, starting in 2024, and by 4.4% each year, starting in 2028. The path beyond 2030 has not been determined yet, but assuming that the 4.4% linear reduction factor persists, it would mean that by 2039 no more emission allowances will be issued (Fig. 3.18).[28]

In addition, carbon capture and storage (CCS) can be especially valuable where balancing the increasing fluctuation in electricity generation introduced by input from renewable sources would require the construction of new gas-fired power plants. CCS technology, coupled with hydrogen-readiness, can make the argument for seeing these plants as a long-term, emission-free investment to secure a stable power supply. With the recent changes

---

[28] A linear reduction factor refers to a percentage of the emissions in the base period, e.g., in 2005. See also Council of the EU (2023). *Revision of the EU emission trading system.*

proposed to the EU ETS, as well and faster reduction of the emissions cap, CCS applications are already on the rise, being encouraged and applied across the EU. Since the EU recognizes CCS as a green technology, it grants it access to European Green Bonds as a funding source.[29]

Despite the advances in deployment and technological maturity, the European Commission does not expect a large presence of CCS in the market before 2030. Rather, it expects it to achieve scale closer to 2035 or 2040, when the carbon price hits at least EUR200/tCO_2.[30] On the other hand, the Network for Greening the Financial System (NGFS) scenario assumes a much quicker rollout, arguing that the technology is mature and ready but lacks committed investment funding.

To achieve a quicker rollout of CCS in industry, and thus an accelerated reduction timeline, the utilities sector can play a critical role by decarbonizing more quickly, so as to allow more allowances to be used for industry, buying time for the latter to deploy CCS more widely. In the case of our decarbonization scenario pathways, in which the impact of CCS technology is attributed to the utilities sector for the NGFS and Ff55 scenarios, a four-year implementation gap would still exist to be compliant with a 1.5 °C warming scenario (Fig. 3.19).[31]

The Ff55 and NGFS pathways contemplate even negative emissions, which is achievable when CCS technology is used in bioenergy power plants.[32] To close the implementation gap, a front-loading of investments would be needed until 2030, amounting to an additional EUR32.6bn per year for power grids and an additional EUR35.8bn per year for power plants. By 2030, to achieve the 1.5 °C goal, the coal exit must be complete and new gas-fired power plants, essential to stabilize the supply of electricity from renewables, must be fitted with CCS. By 2050, the Ff55 legislation expects around 5.4Gt of carbon emissions from the utilities sector, three and a half times the budget

---

[29] Council of the EU (2023). *Revision of the EU emission trading system.*

[30] European Commission (2021). *2030 Climate Plan Impact Assessment.*

[31] In Fig. 3.19, NGFS refers to the Network for Greening the Financial System, PIK for the Potsdam Institute for Climate Impact Research, IIASA for International Institute for Applied Systems Analysis, and PNNL for Pacific Northwest National Laboratory. The Ff55 EU TECH and Ff55EU LIFE scenarios extend the EU 2030 climate target plan for the long-term scenario analysis.

[32] Although bioenergy plants are already considered renewable and carbon-neutral because of their feedstocks, they still use combustion for power generation. Thus, they still technically emit carbon dioxide. Additional CCS applied at their combustion site would result in negative emissions, since the biomass used as feedstock captured CO_2 from the atmosphere while it was growing, and that carbon would now not be returned to the atmosphere but stored in a carbon sink.

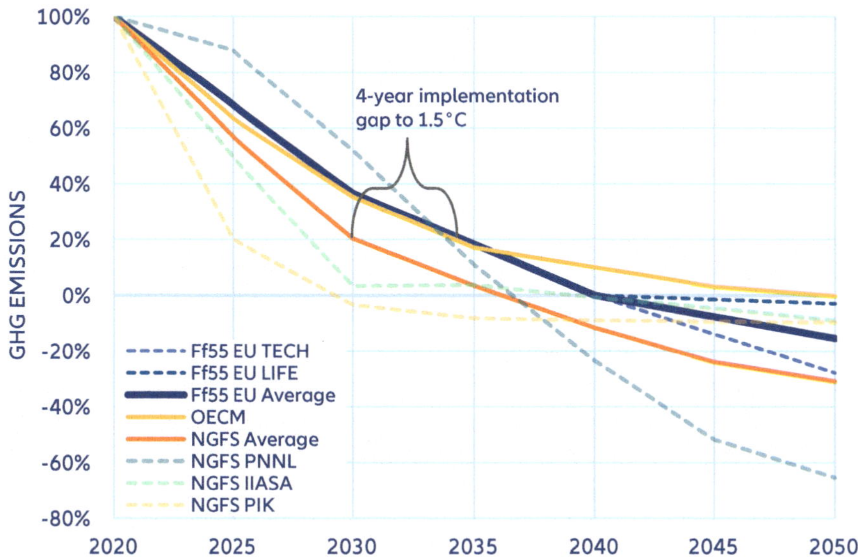

Fig. 3.19 Emissions pathway for utilities: Implementation gap. Source: Allianz Research, NGFS, European Commission 2030 Climate Target Plan

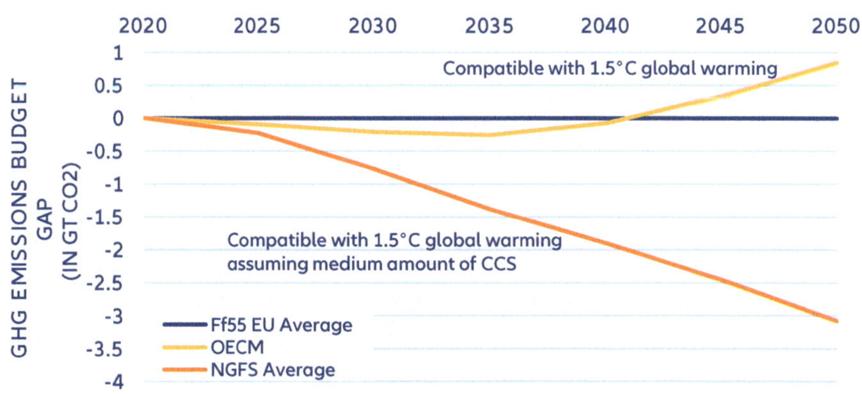

Fig. 3.20 GHG budget gap pathway. Source: Allianz Research, NGFS, European Commission 2030 Climate Target Plan

allocated to utilities by NGFS (1.5Gt) and about 14% of the total remaining carbon budget for the EU—40Gt—for staying below 1.5 °C (NGFS, cumulative PIK path). As Fig. 3.20 shows, this implies that the EU will overshoot the 1.5 °C path by 3Gt until 2050.

Conclusion

Electricity is at the core of the energy transition. Renewables are on the rise and fossil fuels on the wane. Wind and solar energy are cheaper today than their dirty equivalents from coal, gas or oil, and their prices are continuing to drop. But this is not yet enough: the installed capacity of wind and solar must triple by 2030 if the EU is to meet the new demand for electrification across sectors, mostly in transportation and industry. Availability and utilization are the next challenges: "smart" power grids must be ready and digitalized for domestic and transborder trading when weather conditions are less than ideal.

This requires the allocation of sufficient land, which will cause land-use conflicts. The potential solutions to these conflicts, such as rooftop-photovoltaics or agriphotovoltaics, come with additional costs. Furthermore, nearly everywhere the expansion of renewables is exposed to acceptance issues, "not-in-my-backyard" mentalities and lengthy approval processes. In particular, biodiversity and climate protection are often conflicting issues, where in reality they should be viewed and assessed as a joint concept that is not focused on a local assessment but takes account of the larger national and international picture. Decision processes must speed up, and the participation of local citizens and communities in the financial benefits can boost acceptance.

The dependence on dirty coal for supply stabilization must stop by 2030 if a 1.5 °C climate scenario is to be achieved. As coal profitability and CO_2 prices in the EU ETS are directly linked, the best solution would be to reach emission prices in the EU ETS that drive coal-fired power plants completely out of the EU market by 2030, and thus also prevent carbon emission leakage to other EU countries. Research suggests that an increase of the EU ETS price to EUR152 per ton by 2030 would be necessary for this. To achieve the accelerated coal phase-out, an additional EUR131bn is needed across the EU until 2030, with EUR96bn specifically for the coal-6 countries to complete their exits. The gap that coal leaves behind must be filled by an additional 100GW of wind and solar, plus an additional 15GW of new gas-fired power plants. These "hydrogen-ready" gas power plants will be critical to ensure a stable supply of electricity when weather conditions are not ideal, especially during the *Dunkelflaute*, the period when sunshine and wind are absent.

Despite the promising advance that the Fit for 55 proposal would imply for the climate, there is still a five-year implementation gap between the measures proposed and a 1.5 °C warming limit. This gap implies major investment consequences: to stay on the 1.5 °C path, EUR40.8bn per year for power grids and EUR44.7bn per year for power plants would need to be front-loaded until 2030.

The projected emissions reduction is steep, but not steep enough. Therefore, amendments to the EU ETS have been proposed that would result in no emissions certificates being left by 2040, which is 17 years earlier than envisioned under the current yearly reduction rate. This is where CCS technologies can help sectors decarbonize more quickly, but deployment must occur sooner and faster. In order for the Ff55 pathway to be compliant with a 1.5 °C scenario, either CCS technology must be deployed across industry and utilities at a large scale in the next 10 years or, alternatively, the utility sector must decarbonize even faster than proposed to allow for more emissions certificates to be used for industry, buying time to deploy CCS more widely.

A successful transition to climate-friendly electricity generation will, in turn, pave the way to greening up the next pivotal sector: transportation.

Appendix

Load gradients, minimum load and start-up times of power plants of different technologies

Fuel and Technology	Maximum change (during full load)	Minimum load	Start-up time from cold
Hard coal-steam	2–8%/min	20–50%	4–8 h
Lignite-steam	2–8%/min	40–70%	6–15 h
Nuclear-steam	5–10%/min	50–60%	12–25 h
Gas-steam	6–12%/min	ca. 40%	2–5 h
Gas-turbine	10–25%/min	None	ca. 20 min
Gas-combined cycle	4–10%/min	20–40%	1–5 h

Source: European Commission, "Study on the impact assessment for a new Directive mainstreaming deployment of renewable energy and ensuring that the EU meets its 2030 renewable energy target—Task 3.1: Historical assessment of progress made since 2005 in integration of renewable electricity in Europe and first-tier indicators for flexibility"

4

Transportation

The Long and Winding Road

For most of human existence, transportation has been powered primarily by biofuels. Whether on foot, donkey or horse-drawn cart, the primary source of energy was plants—the only exception in the transportation endeavor being sailing boats. Land use had perforce to strike a balance between raising crops for human consumption, or fodder for draft animals. At any rate, it was eminently carbon-neutral: it simply recycled seasonally captured carbon to power motion.

Then first came coal, with the rise of the steam engine, and oil soon afterwards. This changed everything. The new fuel sources not only liberated land for food crops, easing fears of a Malthusian trap and triggering a large and sustained rise in the number of humans on the planet. They also greatly increased the length, strength, and endurance of transportation strides, making it possible to carry people and goods faster and over far longer distances. Not least, they gave people an exhilarating sense of freedom in choosing places to go.

But, most critically, these fuels put an end to carbon-neutrality. Coal, gas and petroleum are hydrocarbons, i.e., organic compounds consisting entirely of hydrogen and carbon, where the carbon they contain had been removed from the atmosphere millions of years ago and buried in the Earth's crust. Today, the combustion of petroleum-derived products is the main source of energy for transportation—and, by reintroducing fossil carbon to the atmosphere, one of the main sources of anthropogenic greenhouse gases. Thus, these fuels can transport us to many places, but not to Net Zero.

L. Subran, M. Zimmer, *Investing in a Changing Climate*, Professional Practice in Governance and Public Organizations, https://doi.org/10.1007/978-3-031-47172-8_4

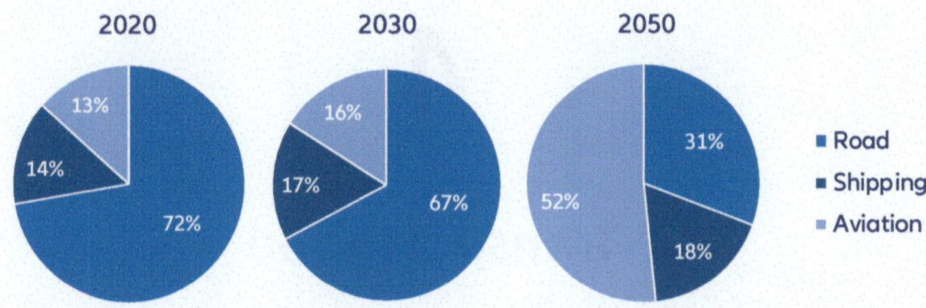

Fig. 4.1 Breakdown of transportation emissions, 2020–2050. Source: Allianz Research

Currently, the transportation sector (including domestic transportation, international shipping and aviation) accounts for almost 30% of the EU's annual carbon emissions, with aviation and shipping on a fairly constant upward trend since 2010—save for a dip during the covid pandemic.[1] The Fit for 55 (Ff55) EU climate legislation aims to wean the transportation sector from its fossil-fuel dependence, emphasizing the use of cleaner, emission-free technologies and alternative fuels. Although there is a relatively clear path for road transportation, with electrification, charging infrastructure and, possibly, green hydrogen for heavy transport, the picture is far less clear when it comes to aviation and shipping. As a consequence, the greenhouse gas emission share of these two industries will sharply increase over time, to almost 50% of all transport-related emissions (Fig. 4.1). More detailed information about the energy balance flows and the relation to other sectors in the economy is included in the appendix.

Comparing the EU Ff55 transport emissions pathway with science-based sector pathways from the Network for Greening the Financial System (NGFS) shows a just- sufficient climate ambition, while the comparison with the One Earth Climate Model (OECM) suggests that the climate ambition still falls short of the emission reductions required for staying within a 1.5 °C global warming range (Fig. 4.2). The necessary pipeline for infrastructure investments and implementation of mitigation measures shows a gap of three years compared to the OECM projection, which relies on lower levels of carbon capture (utilization) and storage (CC(U)S) than the NGFS pathways.

It should be noted that this implies a front-loading of investments, not an increase of the total transition costs: Road transport would need an additional

---

[1] European Environment Agency (EEA). *Greenhouse Gas - Data Viewer*.

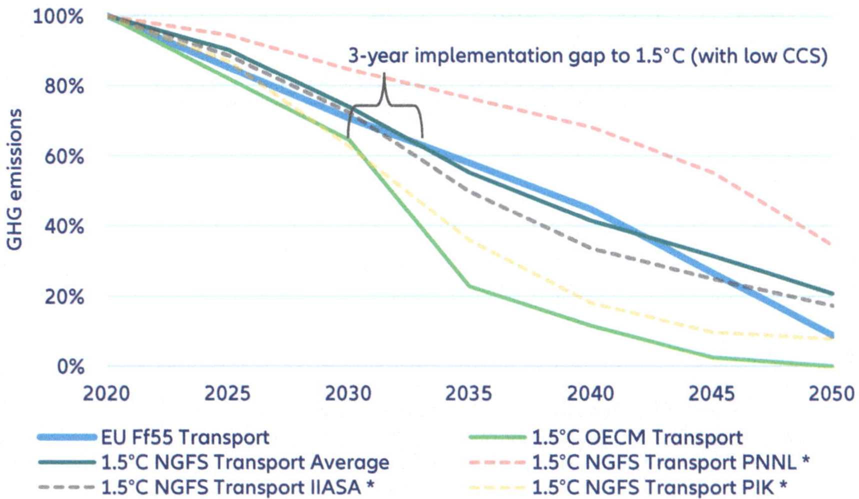

Fig. 4.2 Transportation implementation gap, %. Source: Allianz Research, *PIK is Potsdam Institute for Climate Impact Research, IIASA is International Institute for Applied Systems Analysis, and PNNL is Pacific Northwest National Laboratory

EUR1.1bn per year of front-loading investments until 2030, while aviation would need EUR4.5bn per year. Total EU investment needs for the transportation sector transition are substantial: Until 2050, they average approximately EUR13.4bn per year for electric charging infrastructure and EUR15bn for ramping up sustainable aviation fuel production. In addition, EUR29bn is needed to build up global hydrogen and ammonia infrastructure for the shipping industry, a task that cannot sensibly be addressed by limiting investments to the EU.

Figure 4.3 depicts different paths for total transport sector emissions. Cumulatively, the transport sector under Ff55 will have already emitted 340 Mt. of CO_2 more than the 1.5 °C-consistent OECM pathway by 2030[2] compared to the NGFS scenarios. To put this into perspective, the EU plans to reduce emissions by only 310 Mt. CO_2 using carbon capture technologies until 2030 to achieve the 1.5 °C-consistent emission reduction targets. Carbon capture would thus have to more than double to offset the additional 340 Mt. of CO_2 emissions. However, as the EU generally plans with relatively low carbon-capture activity, it makes sense to compare the evolution of its

---

[2] Negative emissions refer to measures aimed at extracting CO_2 from the air and storing it permanently. A prominent example for this is afforestation.

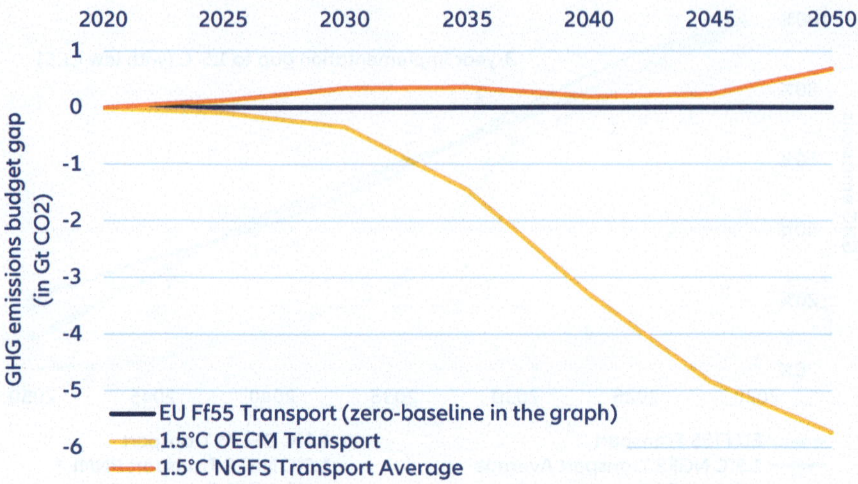

Fig. 4.3 Cumulative emission budget savings in 1.5 °C transport scenarios compared to EU Fit for 55*. Source: European Commission JRC GECO, OECM, Allianz Research. * What is shown is the total deviation of cumulated emissions from the baseline

transportation emissions with other low carbon-capture paths like that from the OECM model. After all, you don't have to capture what you don't emit in the first place.

Road Transportation

Road transportation, which is projected to account for over 70% of overall EU transportation emissions in 2023, shows high potential for decarbonization in the short and medium terms. Technology with zero-emission potential, such as battery-powered electric vehicles, will play a key role in the decarbonization transition. With an eye on 2030 targets, the Ff55 legislation proposes that the average emissions of new cars should decrease by 15% from 2025 to 2029, 55% by the end of 2034, and 100% afterwards. The UK has proposed a similar timeline.[3] Essentially, this means that all new passenger cars and vans on the market would be zero-emission starting in 2035, except for eFuel-only cars. For eFuels, the EU Commission still has to present a Delegated Act that stipulates the extent to which vehicles fueled with eFuels can also fit in this

---

[3] See also Council of the EU. *Infographic - Fit for 55: why the EU is toughening CO2 emission standards for cars and vans.*

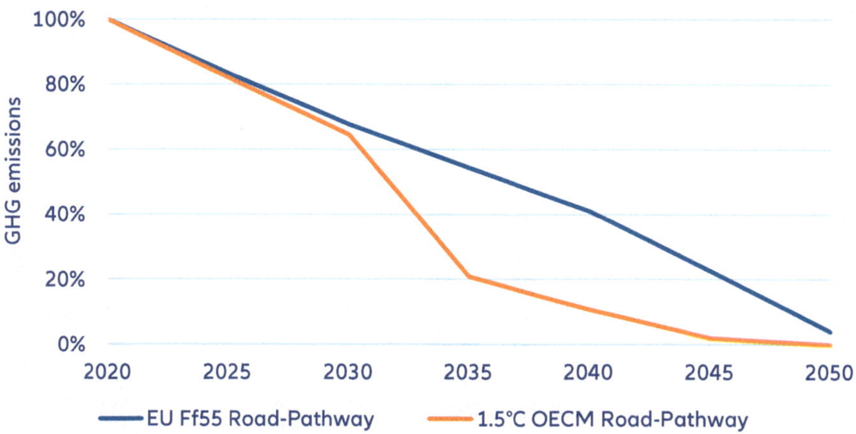

Fig. 4.4 Road transport sector pathways. Sources: European Commission JRC GECO, OECM, Allianz Research

legislation. Alas, ambitious as it sounds, the comparison with the OECM pathway makes clear that the EU will still miss the 1.5 °C target by a wide margin (Fig. 4.4). Staying below 1.5 °C can only be achieved through a massive increase in negative emissions. This can be done either in the form of CC(U)S or, if that is not on the agenda, with a much quicker reduction of emissions, along the lines of the path displayed in the OECM projection, or some combination of both. However, to achieve an ambitious OECM-like path would de facto imply the decommissioning of internal-combustion vehicles long before their usual life cycle ends (typically, in the EU, 10–15 years).

Over the past few years, alternative energy carriers such as electric, hydrogen, natural gas and LPG have developed at vastly different rates and acceptance levels. Since 2014, the electric vehicle market has matured very strongly and is by now the dominant alternative to fossil fuels. Although electric vehicles accounted for 18% of new passenger car registrations in 2021, they still represented only a tiny share of total registered road vehicles in the EU (2.3%).[4] With just under six million electric passenger cars on European roads in 2022,[5] and the average age of a vehicle in the EU being 11.5 years, there are still large gaps to overcome.[6]

---

[4] European Alternative Fuels Observatory (EAFO).

[5] Includes hybrid (plug-in electric vehicles).

[6] Check also EEA (2022). *New registrations of electric cars, EU-27*, EEA (2022). *Newly registered electric cars by country* and for a global comparison, EEA (2019). *Cumulative global fleet of battery electric vehicles (BEV) and plug-in hybrid electric vehicles (PHEV) in different parts of the world.*

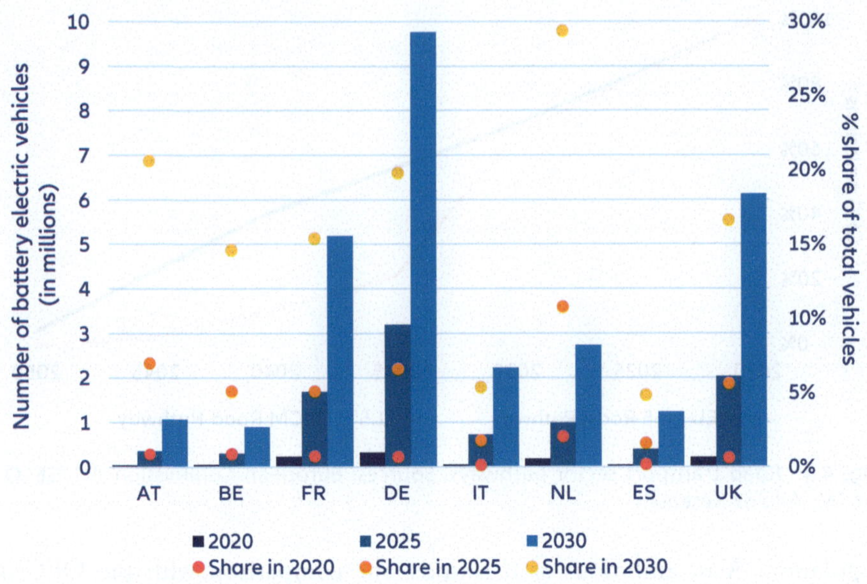

Fig. 4.5 Evolution of electric vehicle numbers in the EU-27, Germany & the UK. Source: 2022 values from IEA global EV data explorer, Allianz Research

For example, the German market had close to two million electric passenger vehicles on the roads in January 2023, with new registrations of BEV vehicles accounting for 18% and BEV + PHEV for 31% of all registrations in 2022.[7] This compares with around 1.5 m electric cars that need to be added each year to meet the target of 14 m electric passenger cars by 2030 and 36 m cars by 2045 (Fig. 4.5), which in turn would imply that 59% of all passenger cars and light commercial vehicles on German roads will have to be battery electric (BEV) or plug-in-hybrid (PHEV) by 2045. By 2050, it is anticipated that 88–99% of all vehicles in the EU will need to use zero- or low-emission technology in order to meet climate neutrality targets. Current projections for the number of BEVs suggest significant growth within Europe until 2030, reaching almost 10 m in Germany alone (Fig. 4.6). However, this is still well below the targeted 14 m expansion.

Apart from an increase in the total number of BEVs, the breakdown of such vehicles by country will also change. Figure 4.6 shows that for the

---

[7] European Alternative Fuels Observatory (EAFO), German Motor Transport Authority (KBA) (2022). *Annual Balance 2022* and Pichlmaier, S. (2022). *Ecological Assessment of Scenarios for the Energy Supply of the German Transport Sector.* (Doctoral dissertation, Technische Universität München). https://nbn-resolving.de/urn/resolver.pl?urn:nbn:de:bvb:91-diss-20221220-1654826-1-3.

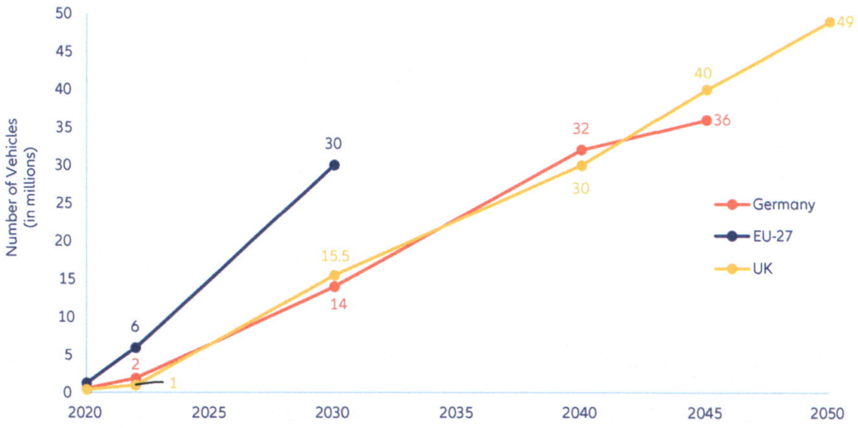

Fig. 4.6 Projected development of BEV vehicles, number and share*. Source: Aalborg University, Allianz Research, * PHEV are not included

selected European countries BEV shares (excluding PHEVs) will increase by 2030 from 4% in Spain to almost 30% in the Netherlands.

The anticipated influx of electric vehicles will approximately triple the transportation sector's electricity demand by 2030 (vs. 2015) in the EU, implying an additional 105 TWh by 2030 and 488 TWh by 2050. This would account for 3.4% of total EU electricity production by 2030 and 9.5% by 2050.[8]

To avoid infrastructure becoming a bottleneck, a large expansion of alternative refueling infrastructure must start now across all EU countries to ensure trans-EU travel. The Ff55 legislation proposes clear targets for infrastructure development for electric and hydrogen refueling stations, stating that the electric charging infrastructure must increase fourfold by 2025 to meet the expected growth of electric fleets. The EU's goal is to have approximately 16 m charging points by 2050 (Fig. 4.7).

Even though the number of electric charging points in the EU roughly doubled between 2020 and 2023 (+191% for AC and 210% for DC charging), the growth is unevenly distributed among the member states.[9] As an example, Germany, France and the Netherlands alone accounted for 63% of all EU recharging infrastructure in 2022. Development also varies depending on the type of charging available. Germany leads in high-power charging

---

[8] Calculated using Electricity Production from JRC (2022). *Global Energy and Climate Outlook 2022 (GECO)*.

[9] *European Alternative Fuels Observatory (EU27)* or alternatively IEA. *Global EV Data Explorer*.

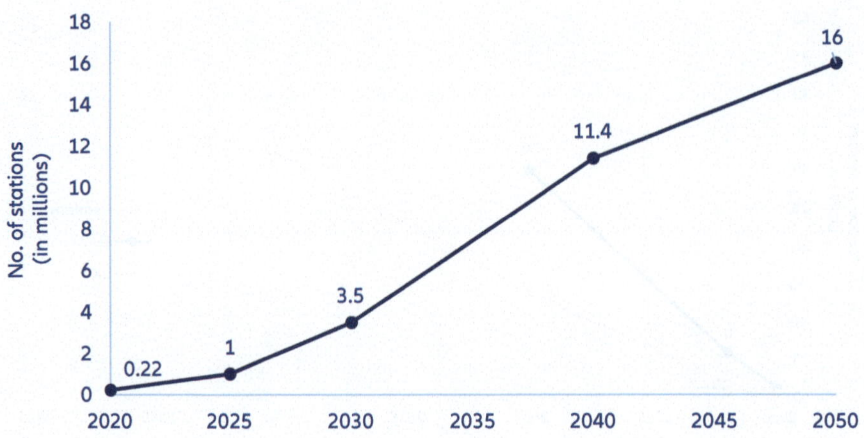

Fig. 4.7 Development of electric charging stations across the EU-27. Source: Allianz Research

(>22 kW) infrastructure among EU countries, with 18.4% vs. the EU average of 12.7%. When it comes to rapid chargers, it is the UK that leads, with around 18.6% of all public charging stations being of this kind, focusing its infrastructure roll-out around increasing access to rapid chargers across its strategic road network.[10] The UK expects to have 2500 rapid-charge points by 2030 and 6000 by 2035.

Currently, the EU has around 500,000 publicly accessible charging points, but trans-EU connectivity is still lagging, as only 7% of the Trans-European Transport Network (TEN-T)[11] was equipped with at least one 150 kW charger every 60 km. The varying charging infrastructure by country will pose a significant challenge for trans-EU travel in the next decade as the electric-vehicle fleet grows (Fig. 4.8).

But with a growing EV market, new business opportunities will emerge for much-needed investment. From 2014 to 2020, Connecting Europe Facility, a funding instrument managed by the EU Commission, awarded an estimated EUR698m for alternative fuel development in road transport, with about EUR343m going towards electric charging projects.,[12] If the proposed Ff55 legislation is implemented, an estimated EUR430bn would be needed

---

[10] *European Alternative Fuels Observatory – Infrastructure* and Bundesnetzagentur (BNA) (05/2023). *Ladeinfrastruktur in Zahlen.*

[11] The Trans-European Transport Network (TEN-T) is the EU's planned network of roads, railways, airports and water infrastructure.

[12] EU Court of Auditors (2021). *Special Report: Infrastructure for charging electric vehicles.*

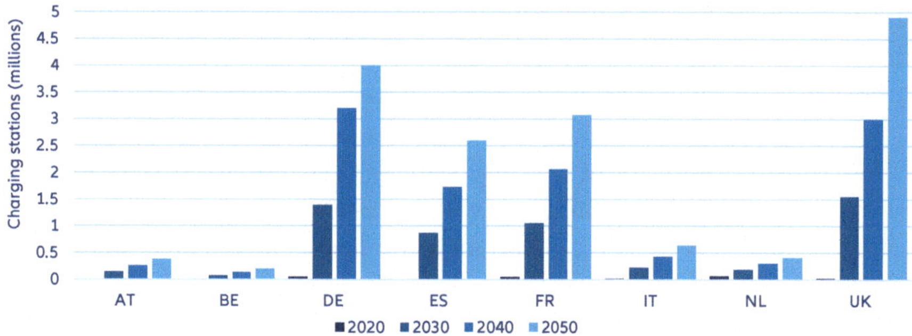

Fig. 4.8 Forecast of electric charging stations by country. Source: Allianz Research

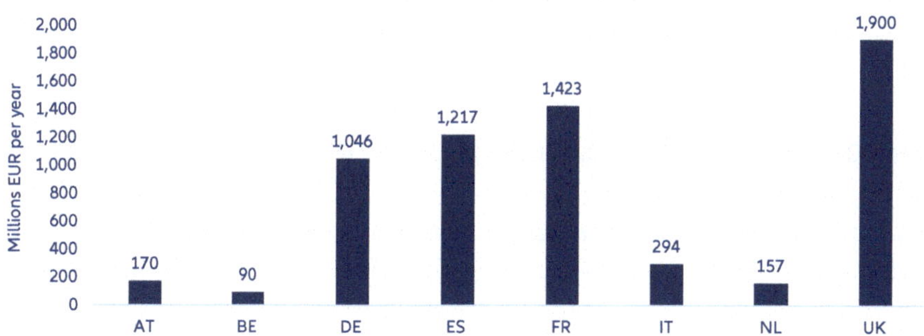

Fig. 4.9 Estimated electricity recharging infrastructure investment needed until 2050, in EUR millions per year [DNV (2020). Energy Transition Outlook]. Source: Allianz Research

between 2021–2050—about EUR13.4bn per year—for recharging infrastructure across the EU-27. Broken down by country, Germany, France and Spain would need to devote the most investment to expanding charging stations to reach the 2050 goals (Fig. 4.9).

In addition, the expansion of the EV market will provide numerous climate-tech opportunities. In Germany alone, an estimated EUR730m per year in start-up financing is needed until 2030 to address gaps in the development and standardization of charging infrastructure and services; closing such gaps will accelerate deployment.[13] With 90% of chargers in Germany being private, growing demand and market pressure will lead to a stable market for charging deployment, especially for residential buildings: approximately two-thirds of households have a garage or parking space, making installation attractive to

---

[13] DENA (2021). *Investing in Net Zero – Assessing Germany's venture capital potential in climate tech until 2030.*

consumers. For Germany especially, with EUR300m in public funding available for SMEs to deploy public recharging stations at their locations, consumer confidence in EVs will continue to grow, especially as purchase grants (of up to EUR6,750 per car) and vehicle tax exemptions continue for EV purchases.[14]

On the other hand, for heavy-duty vehicles, a market for battery technology in truck transport is emerging, with an estimated start-up financing of EUR635m per year needed.,[15] Although the heavy-duty vehicle market has moved more slowly in comparison to that of passenger cars, it is critical for commercial-scale fleet deployment that emission-free technology reaches maturity by 2030. The expansion of public-private investment initiatives, such as Germany's "Electric Mobility Showcase" program, which provided nearly EUR300m to 90 projects, should continue, to provide the market with the boost it needs to meet both demand and the 2030 targets.

Shipping Industry: Uncharted Waters

Maritime transport, a critical component of EU external and internal trade, needs to reduce its CO_2 emissions significantly to meet decarbonization targets. It is expected to account for 13.3% of all EU CO_2 emissions from transport and 3.8% of total EU CO_2 emissions in 2023.[16] To guide the maritime industry towards faster decarbonization, the EU launched the FuelEU Maritime proposal—part of the Fit for 55 package—to foster increased use of renewable and low-carbon fuels in European shipping and ports. Currently, less than 1.2% of the world fleet runs on sustainable fuels.[17]

Based on the 2030 Climate Target Plan modelling, renewable and low-carbon fuels are slated to provide 6–9% of the maritime fuel mix by 2030 and 86–88% by 2050. To achieve this, the draft regulation[18] proposes to limit the greenhouse gas emission intensity of energy used onboard a ship, with increasingly lower limits over time. Starting in 2025, yearly average GHG greenhouse gas emission intensity is proposed to be reduced by 2% from 2020 levels, reducing it further from 2030 by 6% yearly, to reach a 75% reduction by 2050 (Table 4.1). The UK, for its part, is also developing policy

---

[14] German National Center for Charging Infrastructure. *Funding program.*

[15] Investing in Net Zero, DENA (2021) and Energy Transition Outlook, DNV (2020).

[16] EEA (2022). *Member States' greenhouse gas (GHG) emission projections.*

[17] DNV (2022). *Maritime Forecast to 2050.*

[18] Council of the EU (Press Release 03/2023). *FuelEU Maritime initiative: Provisional agreement to decarbonise the maritime sector.*

Table 4.1 Proposed reduction targets for GHG emission intensity of energy for shipping

Target year	Proposed GHG emission intensity reduction target for a ship's onboard energy
2025	2%
2030	6%
2035	14.5%
2040	31%
2045	62%
2050	80%

Source: European Commission

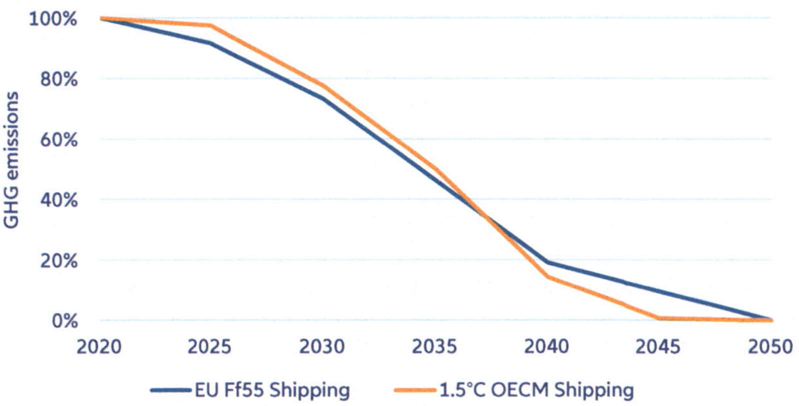

Fig. 4.10 Projected maritime transportation GHG emissions. Sources: European Commission JRC GECO, OECM, Allianz Research

frameworks and instruments to accelerate decarbonization of the maritime sector, aiming to establish, after public consultation, its Course to Zero plan in 2022, setting targets from 2030 onwards.

Alas, these targets fall short of what is needed to achieve the CC(U)S 1.5 ° C path (see Fig. 4.10). However, compared to road transportation, the gap is smaller, particularly until 2030. After that the gap grows wider, so that a more ambitious policy towards sustainable fuels and the retrofitting of existing ships would be required.

In comparison to the aviation industry's proposed fuel blending mandates, the greenhouse gas intensity limits for shipping allow the industry to decarbonize in a more flexible manner, since the technologies required are still evolving. Developing suitable alternative fuels for shipping has turned out challenging due to energy density, technological maturity, commercial readiness, flammability on board, and emissions such as methane and nitrous oxide.

Reviews of fuel options today show very mixed results. Currently, the cleanest readily available alternative is switching to LNG from heavy fuel oil, which could contribute to a 20% reduction in carbon emissions. However, methane slip—an event whereby gaseous methane escapes into the atmosphere—must be controlled. While the EU Commission supports LNG conversion, some environmental groups disagree[19] and believe its benefits would be short-lived.

Promising fuel options for the near future include (limited) biofuels (19–88% emission reduction depending on the feedstock used), which could be blended with fossil fuels, and more promisingly green hydrogen and ammonia, either of which could be used in fuel cells or as a replacement combustion fuel. When produced cleanly from renewable electricity, the potential for emissions reduction is sizable, but unfortunately there is a lack of technological maturity, commercial readiness and refueling (bunkering) availability at ports. Since there are these cost-benefit trade-offs involved in using these alternative fuels for shipping, uncertainty lingers as investors shy away from what they perceive as high-risk investments.

However, action is needed now: ships ordered in the next five years will impact sector emissions for decades to come. On average, shipping vessels have an operating lifetime of around 30 years, with smaller ships (e.g., general cargo) having even longer lifetimes, of some 40 years. Approximately half of all global vessels are more than 15 years old and a third are more than 25 years old.[20] With new fleet renewals scheduled soon, it is critical that zero-emission technology (e.g., hydrogen or synthetic fuel) be ready for operation at least in smaller vessels by 2030.

In a promising trend, Maersk, the world's second-largest container-shipping group, added in June 2023 six vessels with engines able to run on green methanol, a fuel that can help cut shipping's carbon emissions significantly, bringing the company's total order to 25 such vessels. The global orderbook now stands at more than 100 methanol-enabled vessels.[21]

But the most promising alternative fuel for the shipping industry appears to be ammonia (generated using clean hydrogen). However, high investment is still needed for research, development and infrastructure ramp-up. The University Maritime Advisory Services (UMAS) and Energy Transitions Commission estimate that investments of at least USD1trn are needed globally to reduce by 50% the industry's emissions between 2030–2050.[22] For

---

[19] For example, International Council on Clean Transportation and Transport & Environment.

[20] Hydrogen Europe (2021). *How hydrogen can help decarbonize the maritime sector* or alternatively UNCTAD (2022). *Review of Maritime Transport 2022- Chapter 2: Maritime Transport Services* indicating very similar values.

[21] Maersk (Press Release 06/2023). *Maersk orders six methanol powered vessels.*

[22] UMAS & ETC (2020). *Aggregate investment for the decarbonization of the shipping industry.*

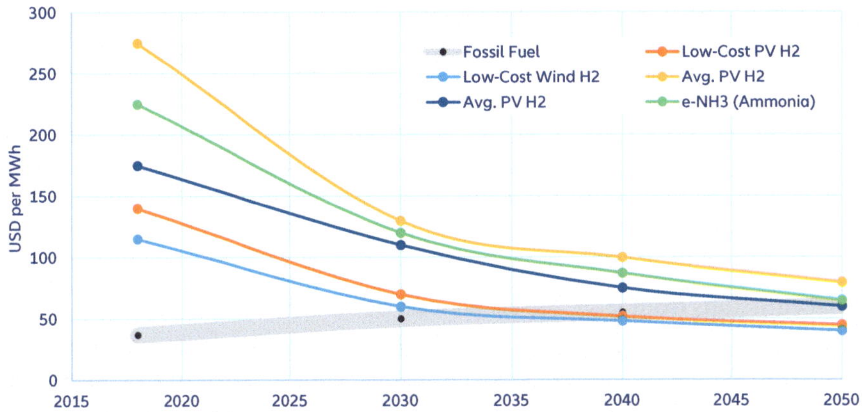

Fig. 4.11 Maritime fuel cost projections. Source: Allianz Research, based on "Navigating the Way to a Renewable Future," IRENA (2019)

complete decarbonization by 2050, which is in line with a 1.5 °C warming scenario, at least USD1.4–1.9trn would be required, with most funding (87%) going towards land-based supply infrastructure such as fuel production, storage and refueling.[23] Half of this 87% would fund hydrogen production and, therefore, also associated PV and wind installations, while the other half would contribute to ammonia synthesis, storage and distribution. On a more granular level, the Sixth Carbon Budget Report[24] produced by the UK's Committee on Climate Change estimates that the UK would need to invest approximately EUR188mn per year until 2035, increasing to EUR412 million per year until 2050, to support the efficiency, electrification and infrastructure adjustments required for zero-carbon ammonia usage.[25]

Decreasing renewable electricity costs will help make alternative fuels more price-competitive against their fossil fuel counterparts. By 2030, hydrogen is projected to be cost-competitive with fossil fuels, while ammonia would be so by 2050[26] (Fig. 4.11). Although hydrogen production is more competitive in the medium term compared to ammonia, the overall capital cost of ammonia is likely to be more attractive, considering that storing it is less challenging (hydrogen requires cryogenic temperatures and high pressure) and that it is commonly used around the world as a fertilizer. As a result, many countries

---

[23] , Global Maritime Forum (2020). *The scale of investment needed to decarbonize international shipping.*

[24] The *Sixth Carbon Budget Report* provides UK ministers with advice on the volume of greenhouse gases that the UK can emit during the period 2033–2027.

[25] See Climate Change Committee (2020). *Sixth Carbon Budget Report.*

[26] IRENA (2019). *Navigating the way to a renewable future: Solutions to decarbonise shipping.*

are familiar with handling and transporting it. In addition, increasing cost-competitiveness of green hydrogen can, thanks to low-cost renewables, help to continue driving down ammonia production costs. On the downside, if ammonia has to be reconverted to hydrogen in the importing country, an energy-intensive cracking process must be undertaken. This is a crucial point, since the importing country as a rule is the one where green energy is scarce.

For the first time, the shipping sector is proposed to be included in the EU ETS, phased in gradually over a three-year period starting in 2023. The operational emissions from large vessels (at least 5000 tons) would include all emissions from voyages within the EU, and 50% of emissions from trips that start or end outside the EU, as well as all emissions while at berth in EU ports. This proposal would cover around 66% of all EU shipping emissions. The inclusion in the EU ETS, together with the FuelEU Maritime proposal, would help incentivize the industry to take steps towards decarbonization and energy-efficiency measures on board vessels. Uncertainty in fuel development, however, could erode the investment potential needed to fund critical research and development in the run-up to 2030.

Aviation: Flightpath Change

In 2019, the aviation industry connected 1.04bn passengers and accounted for 4.1% of the EU's total greenhouse gas emissions.[27] The over 32 million flights carried out in 2022 alone produced roughly 952 Mt. CO_2 (Table 4.2). According to the European Environment Agency (EEA), aviation has been the transport mode with fastest growing emissions compared to 1990 levels.[28] It is likely that aviation emissions will increase further unless decarbonization efforts are ramped up significantly. From a policy perspective, several frameworks have been proposed to direct the aviation industry to address its carbon emissions, including the introduction of fuel-blending mandates for sustainable aviation fuels and the inclusion of aviation into the various emissions trading schemes, such as EU ETS and UK ETS).

However, comparing the EU Ff55 aviation emissions pathway with the low CC(U)S industry pathway from OECM, Ff55 clearly falls short of the required emission reductions to stay within a 1.5 °C global warming range,

---

[27] 2019 figures were chosen to have a pre-Covid baseline. Current passenger numbers for the EU are available at Eurostat (Update 07/2023). *Air passenger transport by reporting country* and current EU aviation GHG emissions EU are available at EEA Datahub (2023). *National emissions reported to the UNFCCC and to the EU Greenhouse Gas Monitoring Mechanism.*

[28] EEA (2022). *Greenhouse gas emissions from transport in Europe.*

Table 4.2 One year in aviation

32,226,001
flights

3.71 Ltr/100 pkm
operational fuel efficiency

111 gCO2e/pkm
operational CO_2e Emissions

952 MtCO2e
net CO_2e Emissions

gCO₂e:	Grams of carbon dioxide equivalent
Ltr:	Liters
MtCO₂e:	Million metric tons of carbon dioxide equivalent
pkm:	Passenger kilometers

Source: Boeing, via *Aviation Week & Space Technology*, Oct. 24–Nov. 5, 2022, p. 61

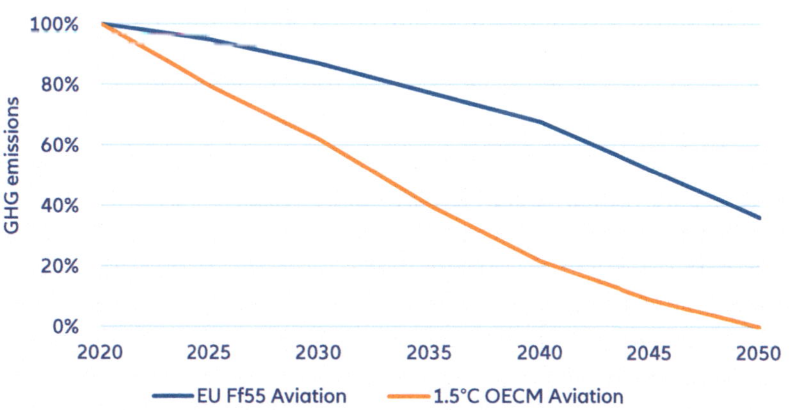

Fig. 4.12 Aviation industry pathway. Source: European Commission JRC GECO, OECM, Allianz Research

particularly after 2030 (Fig. 4.12). Emerging technologies, however, suggest a more ambitious path is achievable. Such a path will not only require a much quicker ramp-up of sustainable aviation fuels, but also an increase in the air fleet's fuel efficiency, both of which will result in higher air fares.

Table 4.3 Proposal for SAF blending mandates for aviation

Target year	Minimum volume shares of sustainable aviation fuel (SAF)
2025	2% SAF
2030	6% SAF, of which at least 1.2% is synthetic fuels[22]
2035	20% SAF, of which at least 5% is synthetic fuels
2040	34% SAF, of which at least 15% is synthetic fuels
2045	42% SAF, of which at least 25% is synthetic fuels
2050	70% SAF, of which at least 35% is synthetic fuels

Sources: European Commission, Allianz Research [European Parliament (Press Release 04/2023). *Fit for 55: Parliament and Council reach deal on greener aviation fuels*]

A promising pathway is the use of sustainable aviation fuel (SAF), which could help reduce emissions by 75%.[29] SAF is a biofuel produced from sustainable feedstocks of biological origins, such as cooking oil, animal waste fat or forestry/agriculture residues. A further option is synthetic fuels, also known as power-to-liquid or e-fuels, which are renewable fuels from non-biological origins, made using renewable electricity. Both SAF and synthetic fuels can replace fossil jet fuel, as they can be directly blended and used to power existing aircraft without any technical modifications, unlike electric and hydrogen-powered aircraft, which, although promising, are not viable alternatives in the short- to medium-term. To create a competitive market for SAF, the ReFuelEU Aviation initiative was proposed in July 2021, which would set fuel-blending targets for the EU-27 starting in 2025 (Table 4.3). It would apply to all commercial air transport flights and set gradually increasing minimum shares of synthetic aviation fuels over time, in line with the EU's climate objectives. In addition, the UK is also planning to consult on a potential UK SAF blending mandate as part of their larger "Jet Zero 2050" plan.

Currently there are seven certified production pathways for advanced SAF biofuels, with varying levels of technological maturity. Perhaps the most readily available and technologically mature fuel is using the Hydroprocessed Esters and Fatty Acids (HEFA) process, which is the only SAF currently being used. This fuel can be blended with kerosene up to 50%, contributing to CO_2 emission savings of 20–69%. Meanwhile, alcohols-to-jet (AtJ) and biomass gasification with Fischer-Tropsch synthesis (Gas+FT) are less mature, but also provide promising CO_2 emissions savings (37–70% and 85–95%, respectively). Because of their well-developed technology, HEFA-pathway biofuels currently have the lowest production costs in comparison to other options, but there may be constraints in the future due to feedstock availability, since

---

[29] European Commission (2021). *Sustainable aviation fuels – ReFuelEU Aviation.*

Table 4.4 Estimated production costs in 2020 for various sustainable aviation fuels and synthetic fuels[a]

Production Route	Fossil Jet Fuel	HEFA	Gas+FT	AtJ	Synthetic Fuel
Estimated production cost in 2020 (k€/tonne)	0.6	0.95–1.14	1.7–2.5	1.9–3.9	1.8–3.5

Sources: European Commission, Allianz Research

[a]Electricity-based SAF is part of *EU H2Global* and will be funded with EUR300m in the first round of tenders. As the funding scheme has been more than quadrupled (from EUR0.9bn to EUR4.4bn), there will also be a further increase in the funding for SAF. Considering a cost gap between synthetic and fuel fossil of EUR 2000 per tonne, 150,000 tonnes could be funded. Furthermore, the European Hydrogen Bank will join forces with H2Global and promote European hydrogen projects. See also WEF (2020). *Clean Skies for Tomorrow Sustainable Aviation Fuels as a Pathway to Net-Zero Aviation* and Federal Ministry for Economic Affairs and Climate Action (2022). *Launch of first auction for H2Global.*

the most suitable feedstocks are in high demand to produce other transportation biofuels (such as for road transport). It is important to note that feed- and food- crop-based fuels are generally disregarded by airlines and not eligible for SAF, because of indirect land-use change and sustainability concerns. Synthetic fuels (or synfuels) that are not sourced from biological origins but rather from renewable electricity have not yet reached commercial availability, with only two processes having been certified for fuel blending—and neither is available on an industrial scale.[30]

The EU countries must accelerate research and production to meet the proposed 2030 targets. In 2020, sustainable aviation fuel accounted for less than 0.05% of total jet fuel use. It is projected to rise to 1–2% in 2027 given the currently planned capacities, mostly due to a lack of price-competitive, mature fuel options, which have been slow to develop.[31] One of the main problems with SAF is limited production and costs. Currently the EU has no plants dedicated to producing sustainable aviation fuel on a commercial scale, which is one of the reasons why production costs are 1.5–6 times higher than fossil jet fuel (Table 4.4). For synthetic fuel, the largest commercial plant recently opened in Germany and was producing approximately eight barrels (1 ton) of synthetic kerosene per day in early 2022.

Although SAF is more expensive, consumers would experience only limited increases in the cost of plane tickets, since fuel costs only make up

---

[30] For more information on certified processes see ICAO. *GFAAF – Conversion processes* as well as processes for co-processing in EASA (2022). *European Aviation Environmental Report.*

[31] See also IEA. *Transport-Aviation.*

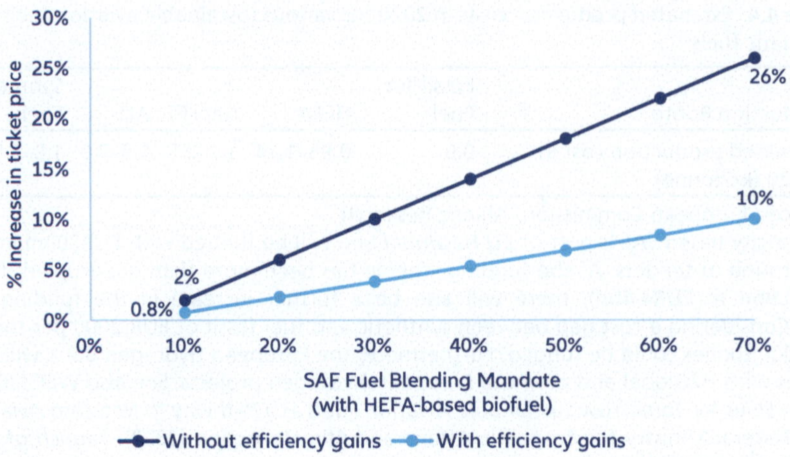

Fig. 4.13 Estimated plane ticket price increase with SAF fuel blending. Source: Allianz Research

approximately 20% of airlines' total costs. Therefore, assuming that SAF is twice as expensive as fossil jet fuel, a 20% fuel-blending mandate[32] (proposed by 2035) would result in a 6% increase in ticket prices (Fig. 4.13). If fuel efficiency gains from newer aircraft are taken into account, ticket prices could increase by 2% with the same 20% blending mandate. By 2050, the proposed fuel-mixing mandate (assuming fuel efficiency gains) would result in an estimated 10% ticket price increase, which is also supported by EU ticket-price estimations in the Ff55 legislation.

HEFA-based SAF is expected to be the main fuel source for blending until 2030, but in the long term the largest share of SAF will come from synthetic fuels.[33] To support the SAF production ramp-up, an estimated 30 plants by 2030 and 250 by 2050 need to be in operation, each ranging from 0.15–0.5 million tons in output per year, with HEFA and Power-to-Liquid (PtL) technologies having the highest share. In Europe, 15 plants are already being planned, largely for HEFA-based fuel production. However, synthetic fuel production needs to start well before 2030 if the targets are to be met. Once a technology is ready for commercial upscaling, it takes approximately three to four years to reach plant operational readiness. Therefore, high investment in technology development is needed upfront to achieve the expected share of synthetic fuels after 2030.

---

[32] Efficiency gains were calculated from the current average fleet consumption of 3.5 liters per passenger per 100 km, while modern efficient aircraft, such as an Airbus A320neo, have an average fuel consumption of 2.3 liters per passenger per 100 km (using data from BDL (2018). *Klimaschutzreport 2018*)

[33] WEF (2021). *Guidelines for a Sustainable Aviation Fuel Blending Mandate, in Europe.*

The London-based Energy Transitions Commission (ETC)—an international think-tank founded in 2015 that aims to accelerate change to low-carbon energy systems—estimates that approximately EUR15bn per year in total capital expenditure (CAPEX) investment is needed in the thirty years from 2020–2050 to ramp up SAF production in the EU.[34] Approximately EUR7bn of the overall total will be required to ramp up HEFA output before 2030, while the largest share of investment (approximately EUR250bn) will be needed to build infrastructure for synthetic fuels, which will play a critical role starting in 2030. Of the synthetic fuel infrastructure, two-thirds of the investment ought to be devoted to building up green hydrogen production.

Public financial support needs to concentrate on research and development of the less-mature pathways (such as synthetic fuels), where approximately EUR30bn over the next 15 years is needed to realize the fuel transition. For example, Germany announced a EUR1bn investment in May 2021[35] to take steps toward a national Power-to-Liquid (PtL) roadmap to cover one-third of its current domestic-flight fuel consumption by 2030.

The UK SAF industry, in turn, could generate between EUR824mn and EUR1.8bn in gross value added per year. To build on this potential market, the UK's Green Fuels-Green Skies competition in 2021–2022 raised around EUR17m to support the development of a first-of-a-kind commercial SAF plant. Another EUR3.5 m has been invested to identify airport infrastructure needs to handle new types of zero-emission aircraft. However, this still falls short of what is needed, which is estimated at an additional investment of EUR460m per year until 2035, followed by EUR672m per year until 2050—just in the UK.[36]

There is some potential for cost-reduction by 2050 through economies of scale, research and development, and a further drop in renewable electricity prices. As an example, the production process for electrolyzers, which split water into hydrogen and oxygen, is currently not automated, even though the technology is ready to be scaled up for commercial production. Apart from this, the cost of green hydrogen can also be reduced through lower operating expenditures coming from increasingly cheaper electricity from wind and photovoltaic systems.

---

[34] WEF (2021). *Guidelines for a Sustainable Aviation Fuel Blending Mandate, in Europe*, which are more ambitious than the investment figures outlined in Fit for 55 (ReFuelEU Sustainable Aviation Fuel).

[35] German Federal Government (2021). *PtL – Roadmap*.

[36] UK Department for Transport (2021). *Jet Zero Consultation – A consultation on our strategy for net zero aviation*.

Change Is in the Air

At first sight, the ire vented by green activists against aviation appears hard to justify. Not only is it economically important, but the carbon intensity of a passenger-kilometer has fallen from around 1.4 kg of CO_2 in 1960 to 0.1 kg in 2018, a drop of about 99%.[37] And it is still falling. As a result, despite the massive increase in flights over the period, aviation accounts for less than 2.5% of global emissions,[38] far below those of cement (8%)[39] or steel (8%).[40] And yet, you do not see crowds brandishing "Shame cement" or "Stop the Steel" banners.

So far, airplanes have become more fuel-efficient through two different means: airframe improvements and engine refinements. Among the former are aerodynamic tweaks such as winglets at the tip of the wings, and lighter materials such as composites for the wing and fuselage structures. Engine advances, in turn, revolve around ever-higher bypass ratios—the proportion of air that goes straight back out of the engine without passing through the engine core, and which provides most of the thrust—and exquisitely optimized cores, the part that compresses the air and mixes it with fuel to power the turbines, which in turn move the big fan at the front and the core's compressor stages.

But however impressively commercial aviation has brought down fuel burn—far beyond what internal combustion motor cars have managed since their invention—it still will not do for these times of climate crisis, in particular because, on current trends, air-passenger numbers are expected to double within the next 20 years. Just like cars, airplanes will have to ditch fossil energy sources if they are truly to become carbon neutral. In addition, they will have to go for more revolutionary airframe shapes than the traditional tube-and-wings.

Boeing, supported by NASA, is exploring a transonic truss-braced wing concept, featuring a slender wing that is 50% longer than that of a current narrowbody and is braced by trusses. This shape alone promises fuel savings of 10% over current similar-sized planes.

Airbus, in turn, is exploring both aerodynamic and propulsion innovations with its ZEROe program, an ambitious undertaking that aims to develop the world's first "zero-emission commercial aircraft" by 2035. It is focusing on three hybrid-hydrogen aircraft: a narrowbody turbofan, a turboprop, and a fat blended-wing one. This last airframe (also being explored by Boeing) is a very wide fuselage that smoothly tapers off to form the wings. In all three, liquid hydrogen is used as fuel for combustion with oxygen and also in fuel cells to create electrical power to complement the gas turbine: hence the "hybrid" in the propulsion system's name. The relevant technologies are being tested on an A380 ZEROe demonstrator, with the aim to achieve a mature readiness level by 2025. Water vapour is produced in both the turbine and the fuel cell which, particularly at the altitudes commercial aviation aircraft fly, causes an increase in

---

[37] Lee et al. (2021). *The contribution of global aviation to anthropogenic climate forcing for 2000 to 2018.* Atmospheric Environment, 244, 117,834. https://doi.org/10.1016/j.atmosenv.2020.117834

[38] D.S Lee et al., op. cit.

[39] Beyond Zero Emissions (2017). *Rethinking Cement.*

[40] McKinsey (2020) based on World Steel Association. *Decarbonization challenge for steel.*

radiative forcing. As with all combustion processes, nitrogen oxides are also produced in the turbines, which can also accelerate the greenhouse gas effect.

Then comes the fuel. Let's leave aside battery-powered aircraft, as they will be limited to air taxis and short-range, small-capacity planes and therefore will play a fairly minor role in reducing flight-related emissions. To power airplane engines, given that the fossil varieties are soon to be no longer socially acceptable, the key concept now is "sustainable aviation fuels", SAF for short (see main text). There are two basic varieties: biofuels, and synthetic fuels that combine clean hydrogen and CO_2, both of which could be used as "drop-in" fuels in existing engines. The problem with biofuels is that their availability is constrained by the shortage of land to produce the feedstocks sustainably, while the SAF produced from discarded cooking oil and animal fats suffers from the limited availability of the feedstock.

That leaves synthetic fuels, and that usually implies hydrogen. If electrolysis is used in places with abundant renewable resources, such as the Atacama Desert in Chile or the deserts in Australia or the Arabian Peninsula, hydrogen could be produced cleanly at low cost.

To turn it into fuel, hydrogen must be combined with carbon monoxide derived from CO_2 obtained either as the by-product of other processes or captured directly from the air (direct air capture, DAC). This is what the HIF Haru Oni eFuels Pilot Plant, in partnership with Siemens and Porsche, among others, is doing in Chilean Patagonia, producing synthetic fuels from electrolysis-based hydrogen and air-captured CO_2.[41] The resulting fuel can be refined into kerosene, diesel for marine transportation and naphtha for use in the chemicals industry.

While carbon-neutral options are still more expensive than fossil jet fuel, rising prices for emissions permits, together with the economic benefits of scaling up, will make the new technologies more competitive. In addition, quotas for SAFs in the EU are forcing the adoption of these fuels.

So, to wrap up, where does it make sense to invest? Bill Gates, for one, has decided that both DAC and SAF are worth a shot and is putting money into them: any amount of SAF produced will find ready buyers. That is the reason why airframers such as Boeing and Airbus, as well as airlines such as United, Lufthansa, Emirates and Etihad, among others, have partnered with SAF producers in the past couple of years. And regardless of whether airlines opt for biogenic SAF or electric propulsion, green hydrogen will play a growing role in the production of so-called e-fuels, among many other uses, making it a worthy target for investment. Makers of electric motors for aviation, finally, are also worth a look, since while not as gigantic as the commercial flight segment, the market for electric aircraft is by no means negligible.

So, whichever way you look at it, when it comes to aviation, change is in the air.

---

[41] IEA. *Direct Air Capture.*

In addition to the proposed new fuel-blending regulation, there are two proposed market-based measures, the EU Emissions Trading Scheme (ETS) and the Carbon Offsetting and Reduction Scheme (CORSIA), that will also support the industry in additional emissions-reduction efforts via carbon pricing and offsetting. For flights within the European Economic Area (EEA), it is proposed that starting in 2024, 25% of the free allocation of emission allowances under the EU ETS would end and instead be auctioned. The share of freely allocated allowances would then decrease linearly and, by 2027, free emission allowances would be completely phased out. For extra-EAA flights, CORSIA would apply.

CORSIA's goal is to ensure that carbon emissions from international aviation do not exceed 2019 levels. For emissions above 2020 levels, each unit of carbon dioxide emitted must be offset by purchasing an approved carbon credit. In other words, the extra emissions must be sequestered or removed from the atmosphere. Given that arguably two-thirds of aviation's climate impact results from non-CO_2 emissions, for every 1 kg of CO_2 emitted, 3 kg of CO_2e emissions would have to be offset.[42] Leaving aside the temporary reduction in aviation emissions during the Covid-19 pandemic, international air travel is expected to become the largest source of carbon-credit demand in the future.[43] While SAF could be help to decrease airline emissions, it is debatable whether these two mechanisms are strong enough to incentivize the industry towards a much-needed fuel transition.

Conclusion

The ride to net-zero is bumpy on the ground, choppy at sea and turbulent up above, fraught everywhere with high costs and uncertainty (air-rides are literally getting bumpier, as climate change is expected to more than double air

---

[42] See Oeko-Institut e.V. (2022). *Fit for purpose? Key issues for the first review of CORSIA* and Transport & Environment (2019). *Why ICAO and Corsia cannot deliver on climate – A threat to Europe's climate ambition.*

[43] This results not only from the problem of fully substituting fossil fuels in aviation, but also from the problem of applying carbon capture and storage (CCS) technologies for residual or unavoidable greenhouse gas emissions from flying. Stationary installations like cement or steel plants, in contrast, are well suited for CCS.

turbulence).[44] These difficulties notwithstanding, large-scale investments are unavoidable in the next 10 years if we are to conduct necessary research and build the commercial-scale production facilities and distribution networks for alternative-fuel technologies.

This requires a willingness to take (high) risks, since most zero-emission solutions and technologies are not yet market-ready and require substantial research and investment funding. Many will not be ready for market deployment until 2030, which means that their contribution to emissions reductions in the next 10 years will be limited. Therefore, available short-term solutions that can help reduce emissions over the next decade must be used in parallel, albeit with a clear understanding of their "bridge" character to avoid lock-ins into unsustainable pathways that are only partially decarbonized. The trade-offs between short-term emissions reductions and a zero-emission future have to be actively managed—very likely the most difficult balancing act of the green transition.

What happens (or fails to happen) in the 2020s will not only largely determine how close the transportation sector comes to achieving its net-zero goal by 2050, but will also have a significant influence on manufacturing—the subject of the next chapter—, which relies on transportation for both its inputs and for distributing its output.

---

[44] See Storer, L. N., Williams, P. D., & Joshi, M. M. (2017). *Global response of clear-air turbulence to climate change.* Geophysical Research Letters, 44, 9976–9984. https://doi.org/10.1002/2017GL074618, Lee et al. (2019). *Increased shear in the North Atlantic upper-level jet stream over the past four decades.* Nature 572, 639–642. https://doi.org/10.1038/s41586-019-1465-z and Prosser et al. (2023). *Evidence for large increases in clear-air turbulence over the past four decades.* Geophysical Research Letters, 50, e2023GL103814. https://doi.org/10.1029/2023GL103814.

5

Greendustry

Developed countries are known as "industrialized economies" for good reason. Industry, alongside trade and services, was the first thing that made them rich. While the top-notch category nowadays is "post-industrial economies", meaning that the service sector generates more wealth than the industrial sector, the way to achieve a "developed" badge usually goes through developing an advanced technological infrastructure. Which means that, even if services overtake industry, industry in general—even heavy industry—will continue to play a key role in the economy.

Unsurprisingly, industry also plays a key role in the battle against global warming. On the one hand, the basic materials it produces—steel, cement, aluminum, chemicals and others—are essential inputs to construction, manufacturing, transportation, infrastructure, agriculture, and consumer goods. On the other hand, many of these industries are energy-intensive and thus emit large volumes of CO_2. Overall, industry accounts for 14% of the EU total emissions.[1]

Some of these emissions fall under the 'hard to abate' category: steel, plastics, ammonia, and cement. The key approaches to tackle these go from innovations in production processes to eliminate fossil CO_2 emissions from production, to developing a more circular economy, i.e., making better use of the materials already produced in order to reduce the need for new production.

---

[1] Material Economics (2019). *Industrial Transformation 2050 - Pathways to Net-Zero Emissions from EU Heavy Industry*

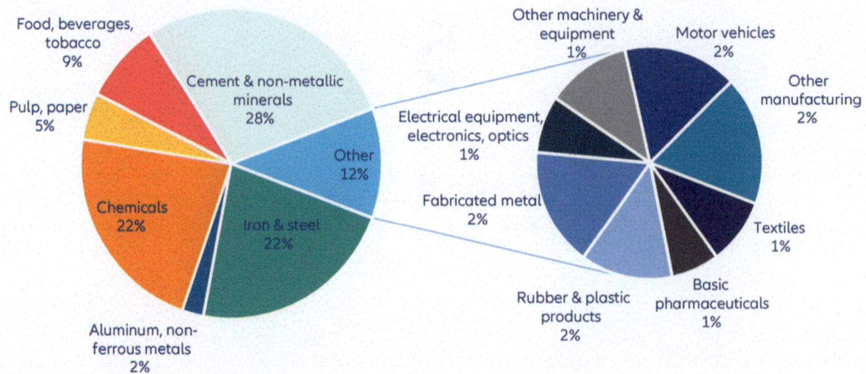

Fig. 5.1 EU-27 + UK industrial CO_2 emissions in 2020. Sources: Eurostat, Allianz Research (excluding emissions from refineries)

Encouragingly, over the past few decades the industry sector has made significant progress towards reducing its emissions and improving energy efficiency. European industry reduced its emissions by 29% by 2010, and by 39% by 2020, compared to 1990 levels. Despite intense international competition, it has managed to adjust its business practices and models to align with the continent's climate and energy goals, maintaining at the same time a viable economic approach.

Nonetheless, the sector is still responsible for 650Mt of CO_2 emissions, which accounts for over 90% of its total direct GHG emissions. The cement, iron & steel, and chemicals industries (Fig. 5.1) are the largest consumers of industrial energy, and thus the largest contributors to CO_2 emissions: they accounted for three-quarters of industrial emissions in the EU27 + UK in 2020.

Furthermore, all three industries produce sizeable process emissions that account for 25% to 50% of the total (Fig. 5.2). This matters because, as mentioned above, industrial process emissions are particularly hard to abate. In fact, even in the net-zero transition scenario, only three-quarters of these emissions are expected to be eliminated in the EU. In contrast, other industries such as food and tobacco; paper, pulp, and print, and nonferrous metals generate mainly indirect and direct emissions (Fig. 5.3), with the former resulting mostly from centrally produced electricity and the latter primarily from heat generation. Therefore, decarbonizing energy and heat generation would automatically reduce a sizable portion of such emissions. For example, nearly 55% of CO_2 emissions in these industries result from the use of centrally produced electricity, primarily from natural gas and coal for low- and medium-temperature heat demand.

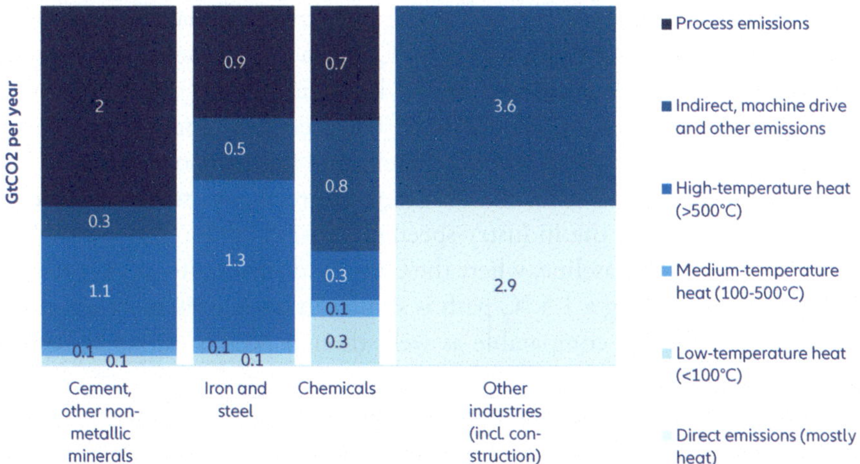

Fig. 5.2 Global CO_2 emissions of different industries by emission source (in $GtCO_2$/yr). Sources: McKinsey [McKinsey (2018). Decarbonization of industrial sectors: the next frontier], Allianz Research

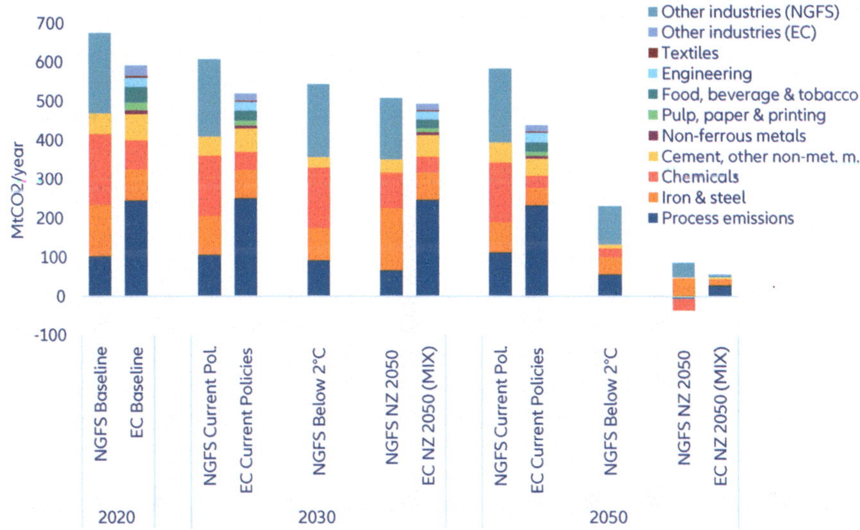

Fig. 5.3 EU industrial CO_2 emissions scenario comparison. Source: NGFS, European Commission JRC GECO, Allianz Research

Figure 5.3 illustrates the gargantuan task of bringing the industrial sector in line with the net-zero pathway. The findings suggest that by 2050, compared to 2020, emissions must be reduced by 92% (and the remaining emissions offset by other sectors like agriculture or through technologies like direct air carbon capture and storage), with some industries even having to generate negative emissions, i.e., capture more CO_2 than they produce. The figure compares the Network for Greening the Financial System (NGFS) projections with the European Commission (EC) assessment for the EU Green Deal. The two sources use different definitions for the boundaries of the industries shown, as well as for the allocation of process emissions and energy emissions. As a result, the industry-specific emissions differ and are slightly higher for the NGFS baseline, where the Other Industries category is broader. The trend for following a 1.5 °C path is similar in both assessments, and net emissions in 2050 are comparable as well, though NGFS explicitly reports negative emissions.

Figure 5.4 shows the development of final energy use in various industries under different scenarios. While the relative composition between industries is not expected to change dramatically, cement, steel and chemicals are expected to have a lower energy-saving potential than the other industries. By 2050, final energy demand in the Current Policies scenario is expected to increase by 14% relative to the 2020 baseline, while it is projected to decrease by 35% in the Net Zero 2050 scenario.

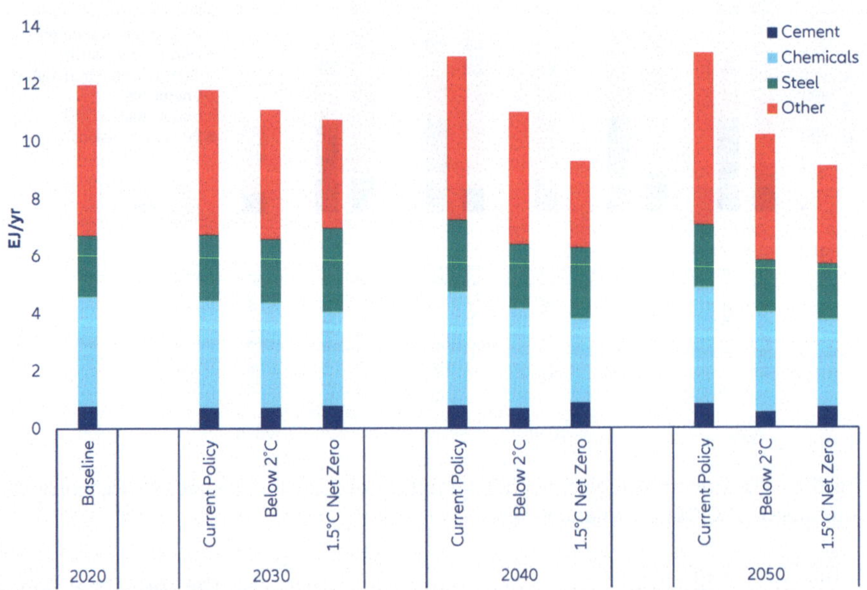

Fig. 5.4 Final energy use by industry and scenario. Source: NGFS, Allianz Research

Options for the Dash to Net-Zero

The different options for decarbonization can be broadly grouped into three sorts: energy efficiency, fossil-fuel substitution through sustainable fuels or electrification, and carbon capture and storage (CCS).

The first two, energy efficiency and electrification, often go hand-in-hand, since they can be two sides of the same coin. Take heat pumps, for instance, one of the main technologies for electrification: not only do they replace fossil energy, but also increase the efficiency of energy use, since they reduce the energy wasted in heating or cooling processes. Whenever cooling is needed, heat will be created as a by-product, with the opposite being true as well.

However, while heat pumps are currently relatively common in residential settings, they are far less established for industrial purposes, although there are large industrial heat pumps (IHP) that can handle thermal processes running on renewable energy or source waste energy from buildings and processes. Applications in the food, paper or chemical industries come to mind.[2] One example for this could be the dairy industry, where milk must be cooled before transport and consumption, while heat is needed for the pasteurization process. Heat pumps can see to it that the heat generated can be put to the best use. However, a significant challenge in many industries is that steam is typically used to transfer heat across a site, resulting in high-temperature system designs. Switching to air or to liquid water requires new pipes, pumps and process designs, which can entail high investment costs and potential disruptions.[3]

There is a hitch, however. While heat pumps increase overall energy efficiency, their net effect depends on where the electricity they run on comes from. Studies show that installing a heat pump that runs on electricity from fossil fuels still has a detrimental net-carbon impact. But as electricity generation becomes greener, every heat pump installed will leverage the positive net effects of this green electricity thanks to its greater efficiency. For example, a

---

[2] IEA (2014). *Application of Industrial Heat Pumps*.

[3] For examples of practical applications of heat pumps in industry, see *U.S. Dept. of Energy (2003). Industrial Heat Pumps for Steam and Fuel Savings*; *IEA (2014)*; European Heat Pump Association (2020a). *Large scale heat pumps in Europe* and EHPA (2020b). *Large scale heat pumps in Europe vol. 2*.

For the methodological approach to emission savings, see FfE (2019). *Small-scale modeling of individual GHG abatement measures in the industry*.

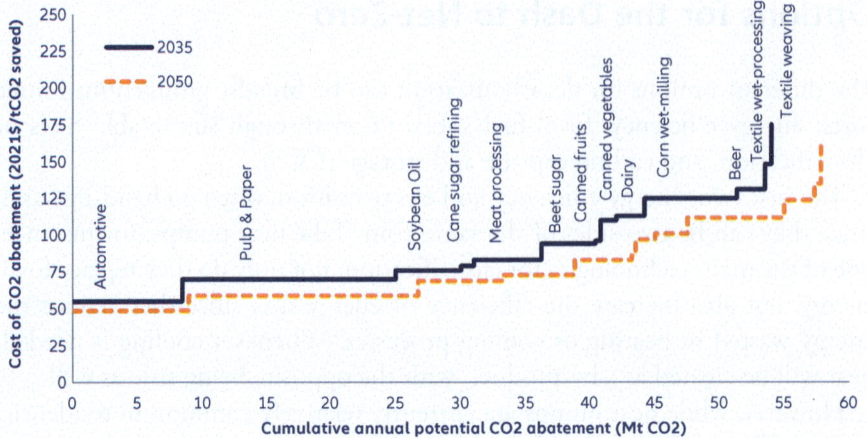

Fig. 5.5 CO_2 abatement potentials through heat pumps in US manufacturing. Sources: Lawrence Berkeley National Laboratory, Allianz Research

heat pump with a COP 3.5[4] emits less CO_2 per kWh of thermal energy compared to natural-gas-condensing boilers when the electricity grid factor is below 740gCO_2/kWh, and is preferable to oil-condensing boilers when the electricity grid factor is below 980gCO_2/kWh.[5]

Hence, the costs of reducing CO_2 emissions using heat pumps will vary widely across industries. A comprehensive study by Zuberi, Hasanbeigi and Morrow analyzes the abatement cost associated with the use of heat pumps in different industries.[6] Their results indicate that with a 100% adoption rate of IHP for hot water and steam-generation systems in 13 industrial processes could reduce annual CO_2 emissions by approximately 17MtCO_2 in the base year 2021. As the decarbonization of electricity grids proceeds, the total CO_2 abatement potential is expected to reach 54.5MtCO_2 per year in 2035 and 57MtCO_2 in 2050, equivalent to 5% of total greenhouse-gas emissions from US manufacturing today (Fig. 5.5). Furthermore, the CO_2 abatement costs are expected to range from USD55 to USD175 per tCO_2 in 2035 (USD50 to USD155 in 2050), depending on the industrial process. Further details on the costs associated with energy savings can be found in Appendix HPUS.

---

[4]COP (Coefficient of Performance) is defined as the relationship between the power (kW) that is drawn out of the heat pump as cooling or heat, and the power (kW) that is supplied to the compressor. A COP of 3.5 reflects the current state of technology.

[5]WBCSD (2020). *Heat pump technologies*.

[6]Lawrence Berkeley National Laboratory (2022). *Electrification of U.S. Manufacturing with Industrial Heat Pumps*.

As good as that sounds, the reality is that regardless of how large the efforts in electrification and other areas of the energy transition are, it is highly unlikely that cumulative carbon emissions between now and 2050 will be consistent with the levels of the Net-Zero 1.5 °C scenario.[7] Industries such as cement and steel have limited potential for emission reduction, since a certain level of CO_2 emissions in their production processes simply cannot be avoided. In other industries, decarbonization efforts are technically possible, but only at a prohibitively high cost. In such industries, Carbon Capture and Utilization or Storage (CCUS) will play a vital role as an economically viable technology that can help them reach their net-zero goals.

Using today's technologies, CO_2 capture rates of over 90% are technically feasible. Carbon capture and storage (CCS) involves capturing the CO_2 from power generation or other industrial activities, transporting it and then storing it in rock formations deep underground. CCUS goes one step further, adding the potential for commercial sale and use of the captured CO_2. Carbon can be captured whenever fossil- or biomass-based fuels are combusted or even before combustion, as in the case of blue or turquoise hydrogen.[8] It can also be applied in the ammonia, iron, steel and cement industries.

The implementation of CCUS has two major use-cases that cut across all industries. Firstly, the most straightforward application: carbon removal. Here, technologies for Direct Air Carbon Capture and Storage (DACCS) and Bioenergy with Carbon Capture and Storage (BECCS) play a major role. Both technologies result in the removal of emissions, so-called "negative emissions", when the captured carbon is permanently stored.

Secondly, CCUS can be applied to capture emissions from industrial processes. The focus here will lie on those industries where emissions cannot be completely removed from the industrial process and other alternative non-CO_2-emitting processes are not available, such as cement, steel or chemicals.

Figures 5.6 and 5.7 show the average and the cumulative CCS investment, respectively, comparing two different sources. While ETC provides a decomposition by CCS technology and industry, as well as additional investment needs in renewable energy to supply power to DACC, the NGFS analysis shows details on the regional split of CCS investments. Around 17% of total

---

[7] ETC (2022). *Carbon Capture, Utilisation and Storage in the Energy Transition: Vital but Limited.*

[8] Blue hydrogen is produced from natural gas, with the resulting CO_2 captured and stored permanently underground. Turquoise hydrogen is made from methane pyrolysis, in which heat separates the hydrogen and leaves solid carbon as a byproduct.

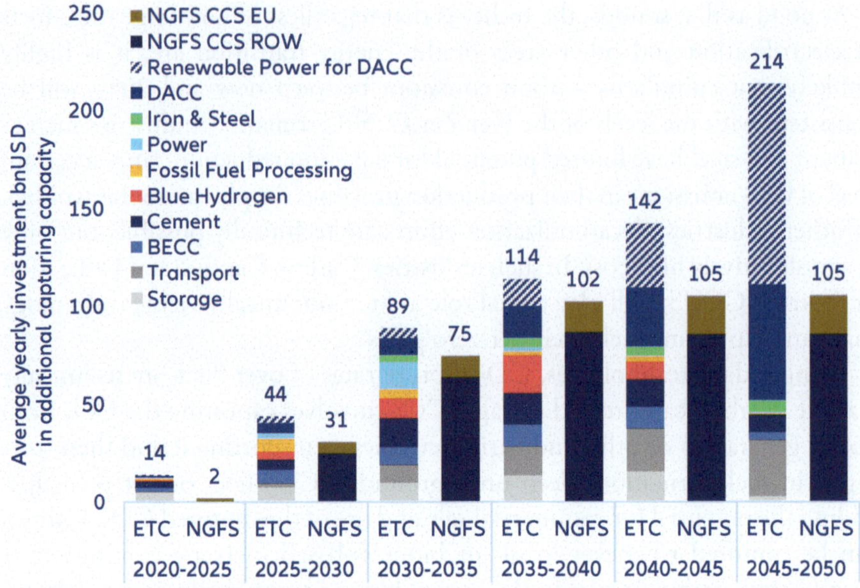

Fig. 5.6   CCS average global investments, USD bn per year. Sources: ETC base scenario, NGFS Net Zero 2050 scenario, Allianz Research

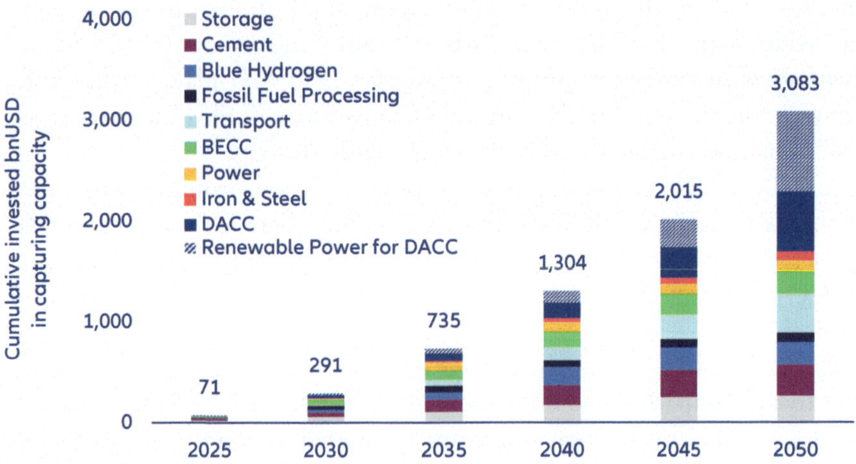

Fig. 5.7   CCS cumulative global investments, USD bn. Sources: ETC base scenario, Allianz Research

investments are currently made in the EU. Notably, investment in Nature-Based Solutions (NBS) are not included, but they are addressed in the section on carbon farming of chap. 6. Given that DACC technologies are usually deployed at or in close vicinity to permanent storage sites, they will need significantly lower investment in transportation (and storage).

Cement and Other Non-metallic Minerals

After water, concrete is the second most-consumed substance in the world.[9] It is also very carbon-intensive: it accounts for 7% of global CO_2 emissions, more than twice as much as global aviation and shipping combined, or more than any country in the world bar China and the US. Without concrete, creating new physical infrastructure would be well-nigh impossible, while the existing one would crumble. Thus, the path to a net-zero global economy unavoidably goes through making concrete—which means basically its main ingredient, cement—clean. While the non-metallic minerals industry consists of a variety of different products such as glass, ceramics, bricks and gypsum, cement and lime production dominate emissions. This includes (1) Process emissions from the chemical reaction that turns limestone into cement; (2) Energy emissions from the generation of the high temperatures needed in cement production, and (3) to a lower extent, emissions from cement transport.

Decarbonizing the cement industry to achieve a net-zero emissions goal by 2050 is challenging mainly due to the difficulty of avoiding its process emissions, in particular those related to the production of clinker,[10] an intermediary product for Portland cement, the world's most common type. The production of clinker is the single most emissions-intensive step in cement fabrication, accounting for 50% of the total. The reason behind this is that the chemical reaction to produce it involves heating limestone to 1450 °C to remove carbon, which then combines with oxygen to form CO_2. The global average clinker-cement ratio is about 0.81, with the balance comprising additives such as gypsum and granulated blast furnace slag from iron production, or fly ash from coal-fired power generation, or natural pozzolana (i.e., volcanic ash).

One possible solution to reduce emissions from clinker production could rely on increasing the proportion of these additives, developing new cement chemistries that reduce the clinker-to-cement ratio, or using clinkers made from alternative raw materials.[11] Other options include recycling of more old concrete from demolished infrastructure.

---

[9] Colin R. Gagg (2014). *Cement and concrete as an engineering material: An historic appraisal and case study analysis.* Engineering Failure Analysis, 40, 114–140. https://doi.org/10.1016/j.engfailanal.2014.02.004

[10] Clinker is produced through calcium carbonate (limestone) calcination, i.e., heating it with other materials (such as clay) to 1450 °C in a kiln. This removes CO_2 from the calcium carbonate to form quicklime, which then chemically combines with the other materials in the mix to form clinker, a hard nodular material that is then ground with a small amount of gypsum into Portland cement.

[11] UN Climate Technology Centre & Network (2010). *Clinker replacement.*

Carbon emissions from generating heat in the kiln could be reduced through a switch from coal to gas, and eventually fully eliminated through heat electrification, or the use of biomass or hydrogen as fuels.

Any remaining CO_2 would need to be captured and stored, which calls for commercializing CCS by 2030. This, in turn, requires infrastructure to transport and store captured CO_2.

Finally, reducing carbon emissions from cement will also require better demand management. Demand growth could be slowed down via greater material efficiency in building design, waste reduction, maximizing the life of buildings and infrastructure, and by promoting some materials circularity.

Given that each of these options would entail significant additional costs, governments are called on to stimulate investment and innovation in these areas by funding R&D and demonstrations, creating demand for near-zero-emission cement or instituting mandatory CO_2 emission-reduction policies, as well as gaining public support for building CO_2 pipelines and storage facilities to enable widespread adoption of CCS.

Cement emissions are being addressed by the EU Emissions Trading System (ETS), while several other countries, including Canada, South Korea and China, have also introduced pricing schemes. Additionally, the EU is developing a carbon border adjustment mechanism for industries, including cement, which aims at limiting carbon leakage and incentivizing stronger emissions measures in foreign countries.[12] Many governments and organizations have also released roadmaps for decarbonizing the cement industry to reach net zero by 2050.[13]

The biggest challenge is to reduce CO_2 emissions while producing enough cement to meet ever-increasing demand; any potential slowdown in Chinese activity, the world's largest cement consumer, could easily be offset by expansions in other markets[14]: while cement demand is projected to stagnate in Europe between 2030 and 2050, it will increase in India, other developing Asian countries and Africa.

---

[12] In the terminology of the European Commission, 'carbon leakage' does not refer only to emissions being emitted in another country instead of the EU, which would not help the global climate ambition: it also refers to the value added and jobs that will be lost in the EU if production gets outsourced to a non-EU country.

[13] In 2022, GCCA published the Concrete-Future-Roadmap, which is summarized in the Appendix GCCA Roadmap. Another roadmap is the IEA Cement Technology Roadmap, which builds on the long-standing collaboration of the IEA with the Cement Sustainability Initiative (CSI) of the World Business Council for Sustainable Development (WBCSD).

[14] IEA (2022). *Tracking report – Cement.*

An extensive analysis of the abatement costs associated with the implementation of the necessary measures, from electrification to CCS, can be conducted using the IndustryPLAN[15] model (Johannsen & Mathiesen 2023). Using a bottom-up approach, the model defines specific measures for the cement industry with adjustable implementation rate parameters to yield results on energy savings and investments for the EU + UK. The aggregate and averaged investments per ton of CO_2 abated for the non-metallic minerals industry (cement, ceramics, and glass) shows a relatively stable relationship at various levels of emission intensity from energy use, with around EUR615/ tCO_2 (Fig. 5.8). In the other industries analyzed, average investment needs will rise more strongly, since marginal cost increases for the last measures to reach zero emissions are typically higher than for the "low-hanging fruit" implemented first. As shown in Fig. 5.8, implementing the measures suggested in the IndustryPLAN model is estimated to result in a decrease of emission intensity from $41.7tCO_2/MJ$ in 2030 to $6.6tCO_2/MJ$ by 2050. Analyzing and aggregating the results published by Material Economics[16] for the cement industry (Fig. 5.9) yields an average global investment of around EUR250/ tCO_2 for an emissions reduction consistent with net-zero targets.[17]

---

[15] IndustryPLAN chooses the decarbonization actions in a bottom-up approach from a merit-order of technology options. More on the background of the technologies and mitigation potentials can be found in Appendix: Industry emission reduction potentials.

[16] Material Economics (2019). *Industrial Transformation 2050 - Pathways to Net-Zero Emissions from EU Heavy Industry.*

[17] Caution: The stated IndustryPLAN numbers refer to reducing the emission intensity of energy use (tCO_2/MJ), while the Material Economics model numbers refer to reducing emissions (% CO_2 total emission reduction). The dots in the Material economics graph show the actual calculated average abatement costs in the model at differing emission reduction levels, while the line shows the ordinary least squares (OLS) estimate derived from the calculated values shown as dots.

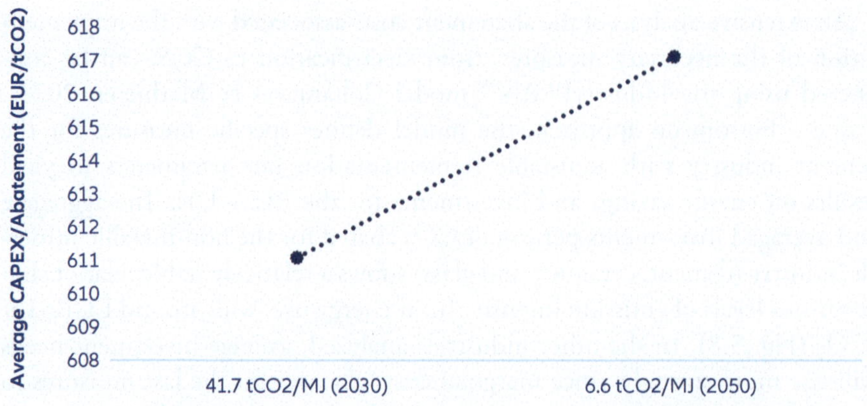

Fig. 5.8 Average investment in the cement/non-metallic minerals industries (EUR/tCO$_2$) to reach net-zero-aligned emission intensity targets. Sources: IndustryPLAN, Allianz Research. Notes: Coverage EU + UK. Non-metallic minerals include cement, ceramics and glass

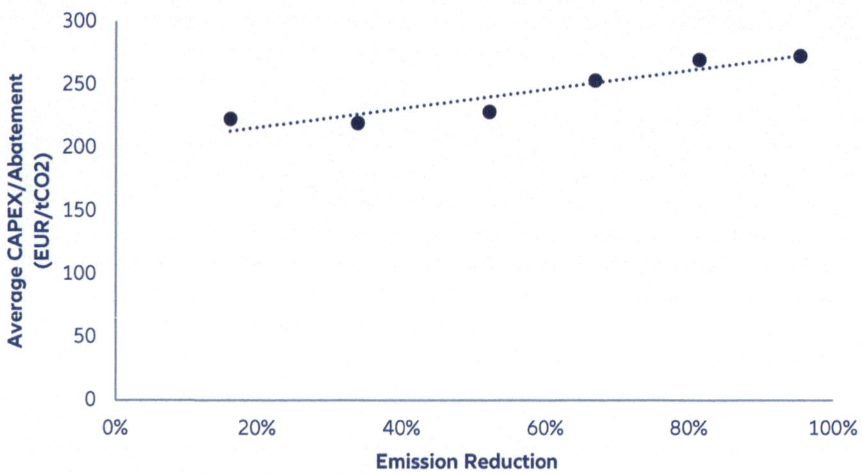

Fig. 5.9 Average cement industry investment (in EUR/tCO$_2$) relative to emission reduction target. Sources: Material Economics, Allianz Research; Note: Coverage is EU

Chemicals

The chemical industry plays a crucial role in the European economy, being integral to major European value chains such as pharmaceuticals, electronics, batteries for electric vehicles, and construction materials. The EU-27 is the world's second-largest producer of chemicals, generating EUR499bn in sales in 2020 and accounting for around 7% of manufacturing output by turnover, which makes it the fourth-largest industry in the EU. The chemical industry employs highly skilled workers and boasts 67% greater labor productivity than the manufacturing sector average.

While chemical production in the EU27 has jumped by 47%, its GHG emissions have decreased by 54% compared to 1990 levels, with energy consumption falling by 21% over the same period. Still, it continues to rank as the third-largest emitter of CO_2 in the EU.[18]

The main drivers of GHG emissions in the EU's chemical industry are the use of natural gas and other fossil fuels as feedstocks for production, along with process-related emissions from ammonia synthesis. The industry requires a significant input of energy in the production of olefins, ammonia and methanol, with ethylene and other olefins dominating both energy consumption and feedstock use. A detailed study by DECHEMA[19] provides a net-zero transition overview for an even broader product range, including chlorine and the aromatics benzene, toluene and xylene.

---

[18] The European Commission developed a *transition pathway for the chemical industry* in the form of a roadmap. It is based on eight building blocks (including competitiveness, funding, infrastructure, skills, the social aspect and more), which were used to sequence key topics against a timeline. The roadmap includes three components: (1) An action-oriented one that groups topics under three cross-cutting themes: collaboration for innovation, clean-energy supply and feedstock diversification. (2) A technology overview and roadmap, based on the SET action plan, its supportive actions and EU initiatives. (3) A regulatory overview, including major R&I initiatives influencing developments in the chemical industry.

[19] DECHEMA (2017). *Low carbon energy and feedstock for the European chemical industry.*

Ammonia

Ammonia production is responsible for around 1% of global emissions and approximately 33% of global chemical Scope 1 emissions, making it the largest carbon-emitting chemical process. The 183Mt of ammonia currently produced annually are used primarily for nitrogen-based fertilizer (70%) and as a chemical feedstock (30%) in various industries such as explosives, mining, construction, plastics, cleaning products and textiles. With the global population set to increase, an additional 44Mt (fertilizer) and 24Mt (feedstock) may be required by 2050.[20] Furthermore, around 50% of greenhouse-gas emissions from ammonia come from Scope 3 emissions, downstream in the application of fertilizers to soil and upstream in fossil-fuel extraction. Beyond its current uses, ammonia has the potential to become the sustainable fuel of choice for the shipping industry.

Ammonia is currently produced by reacting atmospheric nitrogen with hydrogen, using a metal catalyst at high pressure and temperature. Given that hydrogen is still produced using processes that emit large amounts of CO_2, to achieve net-zero ammonia it is crucial to eliminate emissions associated with the production of hydrogen. This can be achieved by using green hydrogen produced through electrolysis with renewable electricity, or blue hydrogen from different variants of steam methane reforming (SMR) or autothermal reforming (ATR) with carbon capture and storage (CCS). Other options are biomass-based hydrogen from gasification of biomass, or biomethane reforming and methane pyrolysis powered by renewable energy. However, today only around 0.1% of the 90 million tons of hydrogen produced every year are of the green variety.

The road to green ammonia, according to ETC, will require sizable investment,[21] ranging from USD4bn to USD12bn for the transitional technologies, while zero-emission ammonia production plants will call for an average investment of USD25bn to USD52bn until 2030. By 2050, an annual average investment of USD59bn to USD109bn will be necessary.

Investment figures for the European chemical industry have been obtained using the IndustryPLAN[22] model (Johannsen et al. 2023). When aggregating (bottom-up) the measure-specific investment costs and emission savings, the analysis suggests that reducing emissions will imply average investments in the European chemicals industry of EUR200 per tCO_2 abated by 2030 and

---

[20] The IndustryPLAN model shows an annual ammonia production of 16.6Mt for the EU in 2020, which is expected to increase only marginally to 17Mt by 2050. However, other scenarios project substantial increases. The MPP ammonia model, for instance, sees a potential increase in global ammonia production from currently 185Mt to 800Mt. The reason for this is that green ammonia might play an important role as a "shipping fuel, as well as in power generation and as a hydrogen vector for long-distance transport for resource-scarce regions." (MPP ammonia report). See also IRENA (2022). *Innovation Outlook – Renewable Ammonia*.

[21] MPP (2022). *Making Net-Zero 1.5°C-Aligned Ammonia Possible*.

[22] IndustryPLAN chooses the decarbonization actions in a bottom-up approach from a merit-order of technology options. More on the background of the technologies and mitigation potentials can be found in Appendix: Industry emission reduction potentials.

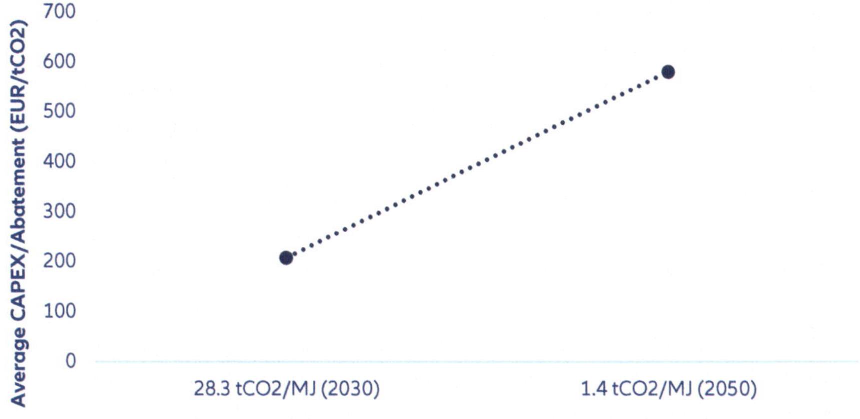

Fig. 5.10 Average investment in the chemical industry (EUR/tCO$_2$) needed to reach displayed emission intensity targets on the path to net zero. Sources: IndustryPLAN, Allianz Research; Notes: Coverage EU + UK

EUR580 per tCO$_2$ abated by 2050, at emission intensities of 28.3tCO$_2$ per MJ and 1.4tCO$_2$ per MJ, respectively (Fig. 5.10). Thus, marginal investment needs to abate an additional ton of CO$_2$ increase as the energy use becomes cleaner. In the case of ammonia, applying the ammonia model developed by the Mission Possible Partnership (MPP) and aggregating the results yields average global investment needs of over EUR700/tCO$_2$ to reduce emissions.[23] According to the MPP findings, most investments must go to green ammonia production (79%) using renewable electricity for hydrogen generation, nitrogen separation and ammonia synthesis (Fig. 5.11).

Iron and Steel

The iron and steel industry is responsible for 2.3Gt, or 7%, of global CO$_2$ emissions. In a business-as-usual scenario, with global steel production expected to increase by 30% by 2050, these emissions could surge to 3.3Gt/y.[24]

---

[23] Caution: The stated IndustryPLAN numbers refer to reducing the emission intensity of energy use (tCO$_2$/MJ) in Europe, while the MPP numbers refer to reducing emissions (%CO$_2$ total emission reduction) globally. The dots in the MMP graph show the actual calculated average abatement costs in the model at differing emission reduction levels, while the line shows the OLS estimate derived from the calculated values shown as dots.

[24] ETC (2020). *Reaching net-zero carbon emissions from steel.*

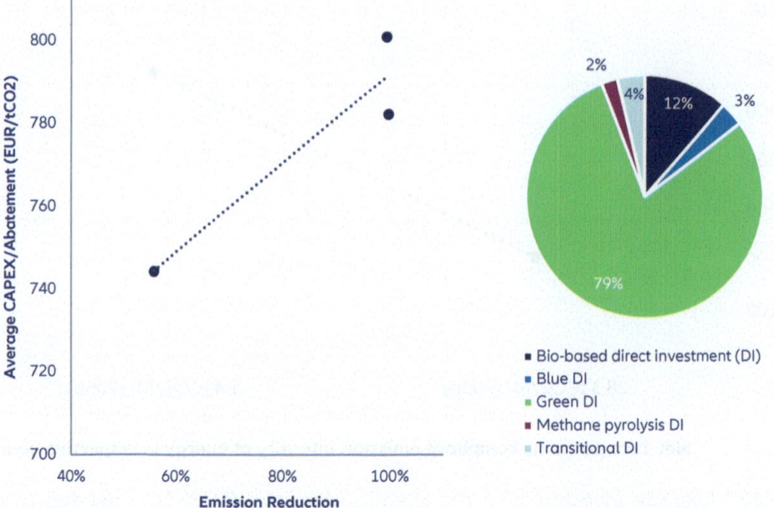

Fig. 5.11 Average ammonia industry investment (in EUR/tCO$_2$) relative to emission reduction target. Sources: MPP [MPP (2022). Report: Making Net-Zero Ammonia Possible – An industry-backed, 1.5°C-aligned transition strategy. Data-Explorer: Ammonia – Pathways to NetZero], Allianz Research; Notes: Coverage is global averages

The production of steel starts with mining iron ore, an energy-intensive activity that generates substantial greenhouse-gas emissions. The ore is then transported to a steel mill, melted and combined with other materials to create steel. This process requires high temperatures that are typically achieved through the burning of fossil fuels. The steel produced is often transported long distances to its destination.

To tackle these emissions, several actions are possible. One is to reduce the total demand for steel. Another is to transition from ore-based (primary) steel to scrap-based (recycled) steel, which could reduce global annual carbon emissions from steel production by 37% by 2050 (and by 52% by 2100) relative to the business-as-usual scenario.[25] Hydrogen-based reduction, while entailing a radical process change, would produce zero-carbon ore-based steel, provided that the hydrogen used is of the green variety. Finally, capturing the carbon emitted could also lower this industry's GHG footprint.

Of the three major technologies used in steel production, Blast Furnace-Basic Oxygen Furnace (BF-BOF) is the most common and the one with the greatest emissions, of about 2.3 t of CO$_2$ per ton of steel. The other two are Direct-Reduced Iron (DRI) using gas as the input, with emissions of 1.1 t of CO$_2$ per ton of steel, and Electric Arc Furnaces (EAF), based on scrap or

---

[25] Material Economics (2018). *The Circular Economy: a powerful force for climate mitigation.*

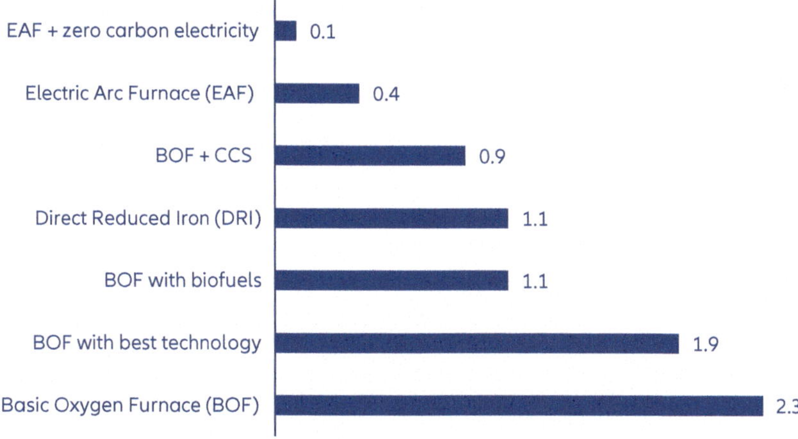

Fig. 5.12 CO_2 intensity of steel production (in tCO_2/t steel). Sources: ETC (2020), Allianz Research

direct-reduced iron and with emissions of about 0.4 t or less, depending on the electricity input (Fig. 5.12). A shift away from the prevalent BF-BOF route towards EAF processes can lead to a significant reduction in the industry's CO_2 emissions. Zero-carbon electricity can reduce nearly all emissions from steel production.

To assess the investment needs for implementing each of the possible emission-reducing measures in the iron and steel industry, we can use the aforementioned IndustryPLAN[26] model (Johannsen & Mathiesen 2023) and the steel model developed by the Mission Possible Partnership (MPP 2022A). Each uses a different bottom-up modeling approach, looking at measure-specific investments for the whole industry (IndustryPLAN) and plant-level technology switches (MPP), respectively. In terms of geographical scope, the IndustryPLAN covers the EU + UK, while the MPP model observes expenditures at a global scale. The resulting investments are shown in Figs. 5.13 and 5.14. In terms of investment efforts necessary for a net-zero transition in the iron and steel industry as a whole, Fig. 5.13 shows aggregate investments relative to achievable abatements for the years 2030 and 2050. Up until 2030, the cost of abating one ton of CO_2 in a net-zero scenario averages EUR117. This increases by 2050 to about $EUR450/tCO_2$. At the same time, emission intensity per energy unit decreases from 58.8 tCO_2/MJ to 4.4 t CO_2/MJ by 2050. For the steel industry in particular, modelling a global net-zero transition using the MPP steel model yields aggregate investment per abatement

---

[26] IndustryPLAN chooses the decarbonization actions in a bottom-up approach from a merit-order of technology options. More on the background of the technologies and mitigation potentials can be found in Appendix: Industry emission reduction potentials.

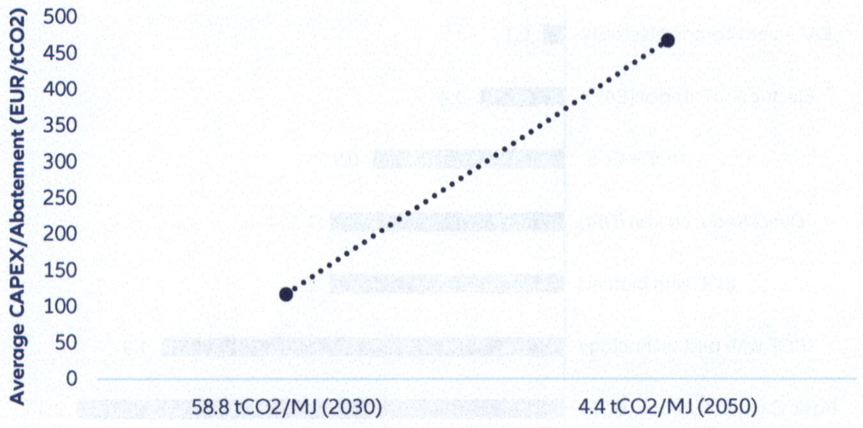

Fig. 5.13 Average investment in the iron and steel industry (EUR/tCO$_2$) needed to reach emission intensity targets displayed on the path to net zero. Sources: IndustryPLAN, Allianz Research; Notes: Coverage is EU + UK

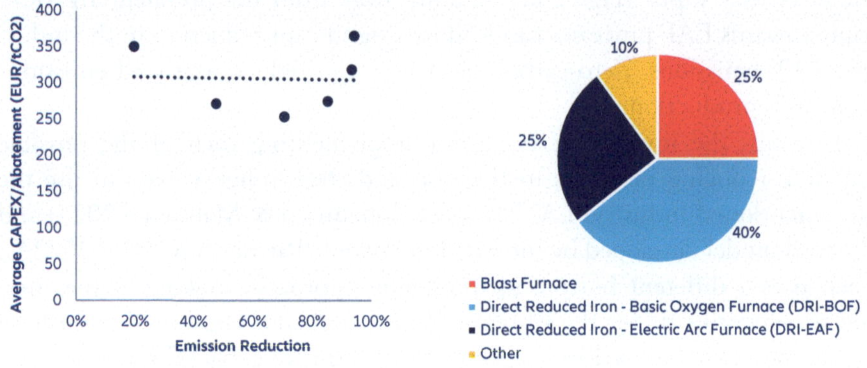

Fig. 5.14 Average steel industry investment (EUR/tCO$_2$) relative to emission reduction targets. Sources: MPP (2022A), Allianz Research, Notes: Coverage is global averages

measure ranging from EUR250/tCO$_2$ to EUR360/tCO$_2$ (Fig. 5.14).[27] Here, the model predicts that most investments (65%) will be used in switching production technologies to less energy- and emission-intensive alternatives such as Direct-Reduced Iron-Electric Arc Furnaces (DRI-EAF) or Direct-Reduced Iron-Basic Oxygen Furnaces (DRI-BOF) (Fig. 5.14).

---

[27] Caution: The IndustryPLAN numbers refer to reducing the emission intensity of energy use (tCO$_2$/MJ) in Europe, while the MPP numbers refer to reducing emissions (%CO$_2$ total emission reduction) globally. The dots in the MMP graph show the actual calculated average abatement costs in the model at differing emission reduction levels, while the line shows the ordinary least squares (OLS) estimate derived from the calculated values shown as dots.

Aluminum and Other Non-ferrous Metals

Non-ferrous metal production is responsible for about 2.4% of industrial emissions in the EU, or 13 Mt. CO_2 annually. Without intervention, emissions are expected to increase by 50% by 2050. China and Southeast Asia, where current producers rely heavily on coal-powered electricity for the smelting process, are expected to meet most of the global demand for aluminum. About 77% of the industry's global CO_2 emissions are generated during smelting, with more than half due to coal-powered electricity usage. Approximately one-third of the industry relies on power from the grid, while the remaining two-thirds use their own power sources.

Recycling aluminum offers a low-carbon alternative. Although recycled aluminum accounts for 30% of the industry's production, it only creates 10% of the industry's emissions. While recycling aluminum requires 95% less energy than primary production, rising demand will make additional primary aluminum production unavoidable. Therefore, the only viable way for this industry to achieve net-zero emissions is to transition to renewable energy sources and CCS technologies.

Transitioning from coal-fired power plants to renewable sources can be challenging for production sites located in areas with limited or unreliable grid power alternatives. In such cases, energy-storage solutions are necessary to ensure that smelters have a constant power supply. Another complication is that plants have long lifespans (30–40 years), making the use of CCS technologies necessary for as long as such plants remain operational. Process emissions, including direct emissions from carbon anodes and fuel combustion during unit processes, account for about 25–30% of this industry's emissions. To reduce these emissions, non-carbon alternatives for the anodes used in the smelting process must be found, and transitioning towards technologies that can provide heat and steam without the use of fossil fuels must be accelerated.

When analyzing the aggregated non-ferrous metals industry branches (such as aluminum, copper, lead, nickel and others) using the IndustryPLAN bottom-up model (Johannsen & Mathiesen 2023), the average investment needs lie in the range of EUR500–600/tCO_2 (Fig. 5.15) and increase slightly as the emission intensity of the energy consumed declines. Notably, emission intensity decreases but does not reach zero, indicating leftover emissions that are not abated directly but could be compensated for by using other means (e.g., CCS). Looking at net-zero-consistent investment strategies for the aluminum industry (Fig. 5.16), the results obtained from the Mission Possible Partnership aluminum model (MPP 2022B) suggest a comparatively wide range of

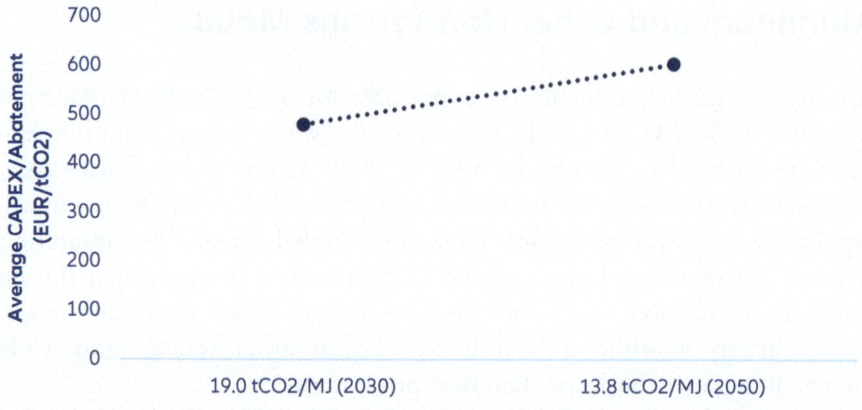

Fig. 5.15 Average investment in the non-ferrous metals sector (EUR/tCO$_2$) needed to reach emission intensity targets displayed on the path to net zero. Sources: IndustryPLAN, Allianz Research; Notes: Coverage Europe + UK. Notes: Non-ferrous metals comprise i.a. aluminium, copper, lead, tin, titanium and zinc, and alloys such as brass

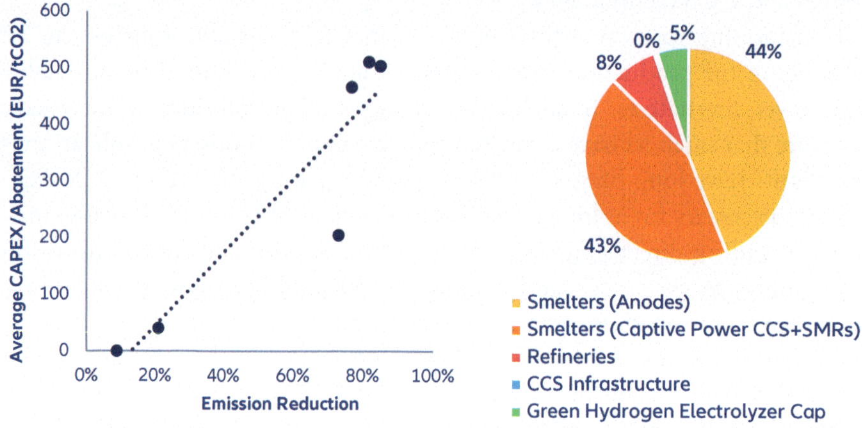

Fig. 5.16 Average aluminum industry investment (in EUR/tCO$_2$) relative to emission reduction targets Sources: MPP (2022B), Allianz Research; Notes: Coverage Global

CAPEX expenditures per unit of abatement, indicating that the "last stretch" of emission-reduction efforts can be substantially more expensive than in the steel industry. In the case of aluminum, the first 10% of emission reductions can be accomplished with an investment of only about EUR50/tCO$_2$, while the reduction of 70–80% compared to 2020 levels will increase the required investment to an average of almost EUR500 per tCO$_2$ abated. For the main investment components, the MPP model finds that 87% of total expenditures are used for implementing new smelter technologies.

Foundries

Foundries contribute to emissions primarily through the energy-intensive process of casting and reforming metals for machinery, vehicle production and other such uses. While their 7.15 $MtCO_2$ in emissions in 2015[28] across the EU28 amounted to only a twentieth of those from iron and steel production, they are still high enough to make them worthy of a serious effort at reduction. According to the IndustryPLAN model, the average investment requirements for an achievable abatement of such emissions come to EUR350/tCO_2 until 2030, increasing to almost EUR450/tCO_2 by 2050 (Fig. 5.17). This would make it possible to reduce foundry emissions by approximately 5.8 Mt., or 81% lower than those in 2015, the base year.

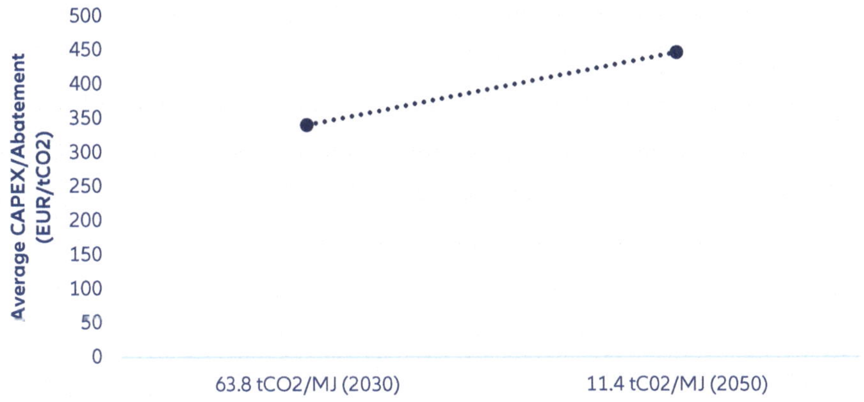

Fig. 5.17 Average investment in foundries (EUR/tCO_2) needed to reach emission intensity targets on the path to net zero. Sources: IndustryPLAN, Allianz Research; Notes: Coverage EU + UK

---

[28] Johannsen, R. M. (Ophavsperson), Mathiesen, B. V. (Ophavsperson) (14 feb. 2023). *IndustryPLAN. VBN. IndustryPLAN_V1(.xlsm).* https://doi.org/10.5278/5fbe12d0-ab4b-4b13-8b33-5c82b527d294

Automotive

The motor vehicles industry as a whole contributes only 2% to direct industry emissions. Yet it receives much attention due to the CO_2 emitted by internal-combustion engines during usage. Such emissions—fuel and exhaust—account for the bulk (65–80%) of a vehicle's emissions throughout its lifespan, and they have already been covered in the Transportation chapter. In this box, the analysis focuses on production-related emissions, which account for only 4–8% of a vehicle's life-cycle emissions.[29]

One of the key approaches to reduce automotive production emissions is the transition to renewable power, which can lead to emission savings of up to 40% (Fig. 5.18).[30] In addition, renewable heat can be used in various manufacturing processes, such as drying in battery-cell production, which can result in a further 20% reduction in emissions. Increasing material and process efficiency, switching the fuels used in production facilities and investing into process innovation can achieve an additional 35% reduction, while the remaining 5% of production process emissions can be tackled using CCS technologies. Despite the necessary investments, rises in consumer costs in the automotive industry are expected to be limited. According to the World Economic Forum (WEF), in a net-zero 2050 scenario, the cost of a car would increase on average by less than 2%.

Figure 5.19 decomposes the energy use and CO_2 emissions associated with the production of a car. To decarbonize the automotive industry, critical processes and footprints must be addressed, including product development, buildings and facilities, manufacturing operations, end-of-life, and supply-chain management.

Power Purchase Agreements (PPAs) enable automobile manufacturers to ensure the use of renewable energy through agreements with their utilities, while the use of industrial internet of things (IIoT) technologies can help optimize operations to reduce energy consumption and waste. To address end-of-life issues, manufacturers are increasingly focusing on designing from the outset for recycling and dismantling, creating the path for an automotive circular economy. This involves using low-carbon resources, materials and assembly; integrating with the energy grid to achieve net-zero carbon emissions across the entire vehicle lifecycle, and disassembling end-of-life vehicles. It also involves recycling batteries and other materials to enable resource recovery and closed material loops, and adopting subscription-based ownership, reuse, and remanufacturing to increase the lifetime of vehicles and components, and ensure efficient vehicle use over time and occupancy. While these initiatives are making progress, there is a long way to go before end-of-life cars can be used as a source of valuable materials. This would make particular sense for secondary aluminum, which is infinitely recyclable and requires just 5% of the energy required to produce primary aluminum.[31]

---

[29] McKinsey (2020). *The zero-carbon car: Abating material emissions is next on the agenda.*

[30] World Economic Forum (2021). *Net-Zero Challenge: The supply chain opportunity.*

[31] World Climate Foundation (2021). *The zero-carbon car: How circular material helps the automotive industry reach their climate targets.*

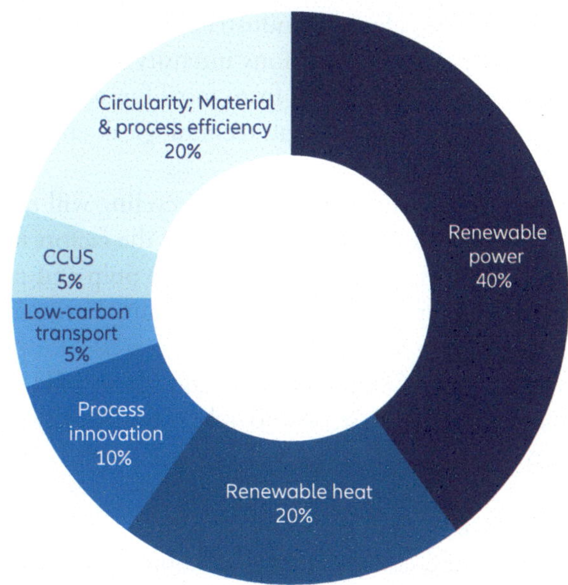

Fig. 5.18 Emission reduction potentials in automotive production. Sources: WEF [World Economic Forum (2021). Net-Zero Challenge: The supply chain opportunity], Allianz Research

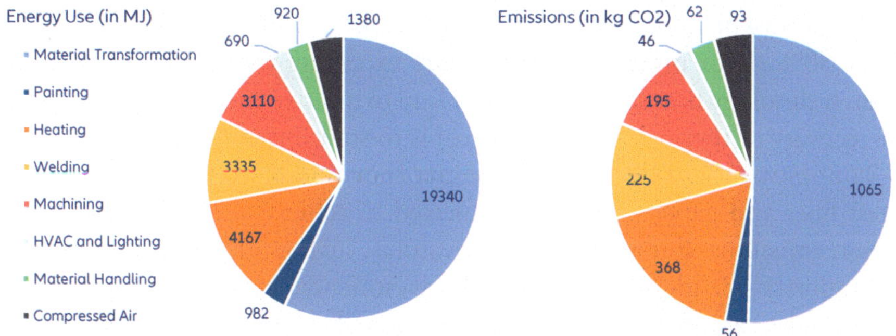

Fig. 5.19 Manufacturing an automobile—energy use and CO_2 emissions. Source: Siemens [Siemens (2022). Decarbonizing practices in the global automotive industry], Allianz Research

Pulp and Paper

The pulp-and-paper industry is highly energy-intensive: it is the third-largest energy user in European industry after chemicals and cement, with 31.8Mt of CO_2 emissions and responsible for about 4% of the EU27's industrial CO_2 emissions in 2021. These very emissions, however, provide the best opportunity for pulp and paper acquiring more climate-friendly credentials.

Since 2018, alas, this industry's energy efficiency has been flat, far from the 4% yearly drop in emissions intensity it requires up to 2030 to be net-zero compliant. A move away from fossil fuels as an energy source and the implementation of meaningful technological change in its processes are long overdue.

Somewhat counterintuitively, recycling will not be an emissions-reduction option for this industry, since one of the factors making it so energy-intensive is the need to evaporate water off the pulp and paper during the drying process. The heat needed for this is obtained primarily from bioenergy, as a by-product of the wood that is the raw material for pulp and paper (this industry is a pioneer in Combined Heat and Power (CHP) generation). Recycled production, in contrast, would rely more heavily on fossil fuels or electricity, since there would be no woody waste to provide the bioenergy needed.[32] The use of bioenergy, while very prevalent in the industry, needs to be increased from 43% in 2021 to 50% by 2030.

As a result, companies are searching for innovative technologies that would allow them to reduce the input of energy and heat in their processes. So far, electricity accounts for a mere 7% of this industry's energy consumption in the EU, overwhelmingly used for heat generation, with fossil fuels providing about 30%. Pilot projects in Europe are exploring the use of superheated steam technology that could potentially lead to massive energy savings through the recovery of thermal energy. The goal is to start introducing it by 2026.

By implementing the various strategies shown in Fig. 5.20, the European forest fiber and paper industry is projected to achieve an 80% reduction in carbon emissions. Energy-efficiency measures, such as process improvements and Industry 4.0 adoption, along with investments in innovative production technologies, are expected to lower emissions by 7Mt of CO_2. Leveraging cogeneration assets (for instance combined heat and power) and participating in the energy market could lead to an additional 2Mt of emission reductions, allowing the industry to better respond to electricity price signals and grid demand, thus adding more demand-side flexibility (lowering demand during high-price hours). The industry's existing use of biomass and gas-based boilers, along with Combined Heat and Power, are estimated to achieve 8Mt of CO_2 emissions reduction thanks to not relying on carbon-intensive fossils. Further emission reductions of 5Mt of CO_2 could be realized through emerging and disruptive technologies. Indirect emissions from purchased electricity are also projected to decrease by 11Mt as European power production continues to decarbonize. Improvements in the transport and logistics chain, including fuel and transport efficiency, infrastructure enhancements, intermodality and the use of alternative fuels such as biogas, advanced biofuels, electricity,

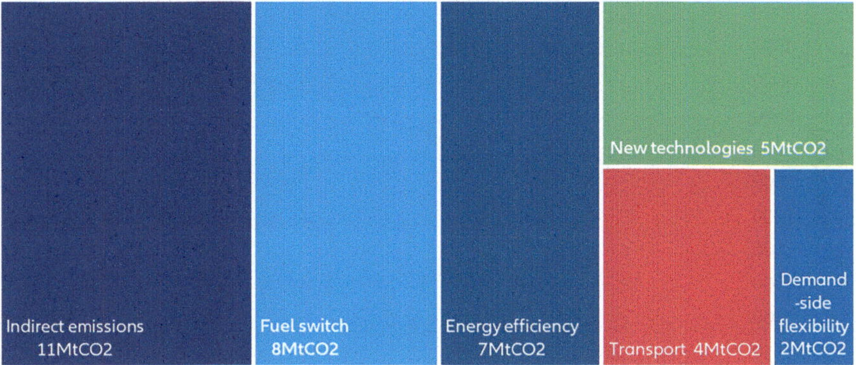

Fig. 5.20 CEPI Decomposition of emission-reduction potential. Sources: CEPI [Confederation of European Paper Industries (2017). Investing in Europe for Industry Transformation – 2050 Roadmap to a low-carbon bioeconomy], Allianz Research

or fuel cells, could contribute with an additional 4Mt of CO_2 emissions reductions.

For the pulp and paper industry, the transition to net-zero raises further challenges. According to the United Nation's Food and Agriculture Organization (FAO), the industry is the largest user of virgin wood, accounting for about 14% of total wood consumption.[33] As soon as other industries (i.e., steel, cement, plastics or ammonia) transition their energy sourcing from fossil fuels to low-carbon electricity and biomass, biomass as feedstock will become a highly sought-after commodity. Next to more common uses of biomass (i.e., biofuel production), non-fossil carbon will be needed to produce petrochemicals and plastics, or as a reducing agent in the steel industry.[34]

Among the industries covered in the IndustryPLAN[35] project,[36] the pulp and paper industry shows the highest average investment needs for CO_2

---

[32] IEA (2022). *Pulp and Paper.*

[33] Del Rio et al. (2022). *Decarbonizing the pulp and paper industry: A critical and systematic review of socio-technical developments and policy options.* Renewable and Sustainable Energy Reviews, 167, 112,706. https://doi.org/10.1016/j.rser.2022.112706

[34] Material Economics (2019). *Industrial Transformation 2050 - Pathways to Net-Zero Emissions from EU Heavy Industry.*

[35] IndustryPLAN chooses the decarbonization actions in a bottom-up approach from a merit-order of technology options. More on the background of the technologies and mitigation potentials can be found in Appendix: Industry emission reduction potentials.

[36] Johannsen, R. M. (Ophavsperson), Mathiesen, B. V. (Ophavsperson) (14 feb. 2023). *IndustryPLAN. VBN. IndustryPLAN_V1(.xlsm).* https://doi.org/10.5278/5fbe12d0-ab4b-4b13-8b33-5c82b527d294

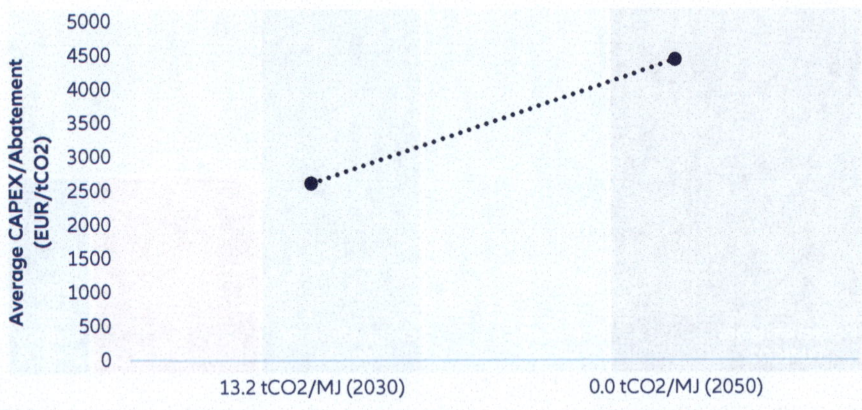

Fig. 5.21 Average investment in the pulp and paper industry (EUR/tCO$_2$) needed to reach emission intensity targets on the path to net zero. Sources: IndustryPLAN, Allianz Research. Notes: Coverage EU + UK

abatement in a net-zero scenario, with over EUR2500/tCO$_2$ (Fig. 5.21). Even though its emission intensity of energy is already initially lower than that of other energy-intense industry branches—making remaining emission reductions potentially more costly—this does not completely explain the substantial cost differences. When looking at the type of investments that need to be made (Fig. 5.22), it is clear that most investment flows are expected to go into electrifying production processes and the adoption of innovative production measures. In the electrification process, the major component adding to costs is the use of high-temperature heat pumps in the production of mechanical pulp or graphic paper.

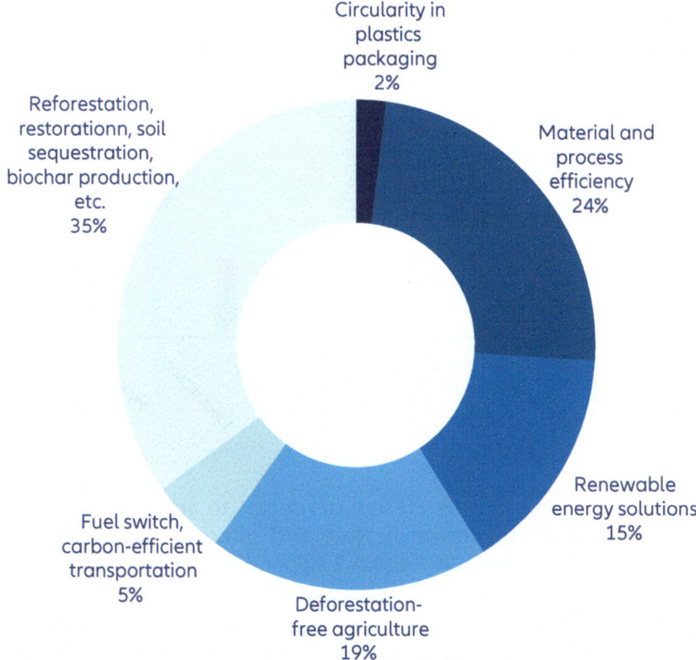

Fig. 5.22 Emission-saving potential in the food industry. Source: WEF [World Economic Forum (2021). Net-Zero Challenge: The supply chain opportunity], Allianz Research

Food, Drink and Tobacco

Climate change discussions often center around major emissions sources such as power generation, transportation and factories, while everyday activities like food and drink consumption receive much less attention. As it turns out, the food value chain, which includes farming, manufacturing, production and transport, is estimated to account for no less than 30% of total carbon emissions in the EU. Focusing on the food production and packaging processes within the industry (and ignoring the rest of the value chain), food, drink and tobacco account for 9% of total EU industry CO_2 emissions (Fig. 5.1). Within the EU, food and drink manufacturing contributes 11% of the total agrifood value chain emissions. Most emissions come from energy use, with 62% consumed as heat and 38% as power from the grid. A significant amount of electricity is dedicated to cooling, higher than what is

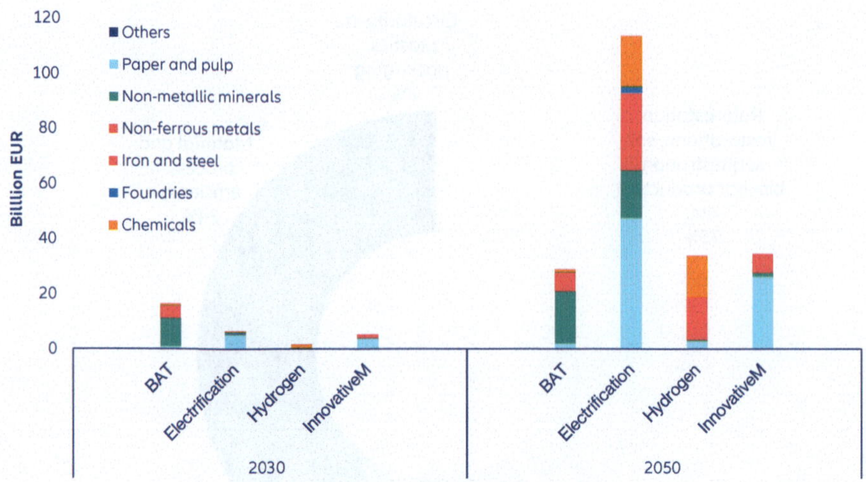

Fig. 5.23 Investment needs in the EU industry sector to achieve net-zero emissions. Sources: IndustryPLAN, Allianz Research. Note: BAT refers to best available technologies. Includes EU + UK. See Tables A.1 and A.2 Appendix for a decomposition of investments by country

typically observed in other manufacturing industries. Some industrial processes and applications require high temperatures that are very difficult to produce without sizable CO_2 emissions; this is one of the most significant challenges for this industry.

The challenge for addressing the bulk of emissions, especially for SMEs, lies in the complexity of value chains that involve numerous actors that can be very far flung. This does not only include the industry at large, but also policymakers enacting regulatory frameworks, society adjusting its behavior, and financial institutions providing the means for transition-fostering investments. A recently published industry roadmap aims to address these challenges and help the sector reach decarbonization goals,[37] turning in the process the European food system into a global standard for sustainability. The EU strategy is to reward farmers, fishers and other food-chain operators who adopt sustainable practices, and to facilitate the transition for others and create more opportunities for their businesses.

According to the WEF, several measures can be taken to decarbonize the agrifood value chain (Fig. 5.23). Packaging made from recycled plastics can result in a reduction of about 2% of emissions. Material- and

---

[37] FoodDrinkEurope (2021). *Decarbonising the food and drink industry.*

process-efficiency improvements can lead to a reduction of approximately 25%. Transitioning to renewable power sources can help to reduce emissions by a further 15%. Additionally, implementing deforestation-free agriculture practices can lead to a reduction of around 20%. Switching to more carbon-efficient transport fuels can contribute to a 5% reduction in emissions. However, it is important to note that approximately 35% of emissions in the food industry are inherent to agriculture and cannot be fully eliminated through these measures alone. Additional steps, such as reforestation, restoration of mangroves and peatland, soil sequestration and bio-charcoal production, will be necessary to fully address these emissions.[38]

ICT Sector

Unlike oil and gas, we cannot even imagine doing without today's information and communications technologies (ICT). Broadband access has practically become a human right, the internet of things is fast turning into our planet's nervous system, cryptocurrencies keep speculators, central bankers and climate activists awake at night, and artificial intelligence is growing by leaps and bounds, permeating ever more areas of daily life.

With electricity powering all these technologies, how this electricity is produced plays a pivotal role in the ICT-related carbon footprint, a footprint that is by no means negligible. The Bitcoin and Ethereum cryptocurrencies alone consume up to 240 terawatt-hours annually, an amount that accounts for around 0.9% of annual global electricity usage and exceeds Australia's yearly electricity consumption.[39]

Add to this the world's cloud computing infrastructure, datacenters, transmission and broadcasting gear, the systems powering artificial intelligence outfits and the myriad devices in our pockets, wrists, ears and cars and appliances and we are talking about an estimated share of global GHG emissions ranging from 1.8 to 2.8% in 2020.[40] That is comparable to global aviation's GHG output.

---

[38] Soil sequestration refers to practices that increase the carbon content—and thus carbon sequestered—in soil. Biochar—a form of bio-charcoal—is used as a form of nature-based carbon removal in which biochar is produced through pyrolysis of residual biomass, followed by the subsequent application of the resulting biochar to soils or durable materials (e.g., cement, tar). For more information, see Chap. 6 agriculture.

[39] Until now, the right to update the blockchain databases that underlie such assets was earned by solving exceptionally difficult mathematical puzzles; the winners (called "miners") were rewarded with freshly minted cryptocurrency. This process, known as "proof-of-work", requires massive amounts of computing power, and therefore energy. So much so, in fact, that China banned crypto mining operations in its territory (which, incidentally, just drove the miners underground: China is still the No. 2 Bitcoin miner in the world after the US).

See also: The White House (2022). *Climate and Energy Implications of Crypto-Assets in the United States.*

[40] Freitag et al. (2021). *The real climate and transformative impact of ICT: A critique of estimates, trends, and regulations.* Patterns, 2(9). https://doi.org/10.1016/j.patter.2021.100340

Fortunately, the process that makes crypto assets so energy-intensive is undergoing a fundamental change. Ethereum, instead of demanding "proof-of-work" as cryptographic proof, has switched to "proof-of-stake" as a climate-friendlier alternative.[41] The new system, apart from being far more secure, requires 99.9% less energy to operate.

What about the rest of the ICT sector? The first step has been to set up some goals. The International Telecommunication Union (ITU), an agency of the UN, has defined a standard that aims at a 45% reduction by 2030 and net zero by 2050. However, given that the 1.5 °C goal is presented as a recommendation, ICT industries are not bound to comply with this voluntary standard. In practice, this means that the trajectory will most likely lie between the business-as-usual and 1.5 °C scenarios.

The definition of the ICT sector's carbon footprint has two components: embodied emissions and operational emissions. The first cover the emissions originating from manufacturing and installation of the equipment and appliances. Operational emissions, in turn, stem from the use of these networks and devices, primarily electricity consumption and related emissions from the global electricity mix. Embodied emissions account for roughly 30% of the total carbon footprint, while operational emissions account for the remaining 70%.[42]

Consumer behavior will most definitely not change drastically towards a reduction in the use of electronic devices in the future—rather the opposite, in fact. Thus, the way to reduce emissions is to increase the share of renewable electricity in the mix. The remaining emissions could be brought down by optimizing the product life cycle, such as by reassessing material selection, design choices, manufacturing and transportation.

Conclusions

Summarizing, the investment needs for sticking to a net-zero-compliant path are substantial but achievable, as the following figure shows. The appendices that follow provide more detail in breakdowns by industry and by country.

Energy efficiency, fossil-fuel substitution through sustainable fuels or electrification, and carbon capture and storage (CCS) are the three main options for decarbonizing industry.

There are, however, major differences within the industrial sector: the hard-to-abate group, which includes iron and steel, aluminum, chemicals and plastics, and the easier group, which is all the rest, including vehicle manufacturing, pulp and paper, foundries, non-ferrous metals, ICT, and the food value chain, among others.

---

[41] Forkast (2022). *China banned Bitcoin mining and became world's No.2 Bitcoin miner.*
[42] Malmodin, J. (2020). *The ICT Sector's Carbon Footprint. Presentation at the techUK Conference in London Tech Week on 'Decarbonising Data'.*

For the first group, which accounts for the lion's share of industrial emissions, the long operational life of mills and manufacturing complexes poses an additional hindrance, since it may deter investors, but with the right combination of incentives and support, the goal is definitely achievable.

For the second group, replacing fossil-generated electricity and heat with the sustainable variety will suffice. Industrial heat pumps, although still fairly rare despite their availability, should make a significant contribution here. Any remaining emissions will have to be dealt with by resorting to carbon capture and storage.

In any case, as with all other sectors in the economy, a massive increase in the availability of renewable energy is *sine qua non*.

Appendices

Decomposition of Investments by Country

Table A.1 Cumulative investment per country for EU27 + UK until 2050 (in EUR million)

Country\ Industry	Chemicals	Foundries	Iron and steel	Non-ferrous metals	Non-metallic minerals	Paper and pulp
Austria	909.0	54.5	3685.2	19.4	1029.4	4525.0
Belgium	3077.9	14.3	2659.7	19.4	1298.6	1787.3
Bulgaria	402.4	7.6	26.0	19.4	571.5	316.4
Croatia	353.8	12.5	4.6	13.2	496.9	158.7
Cyprus	0.0	0.0	0.0	0.0	159.9	0.0
Czechia	239.8	77.6	2539.1	19.4	805.5	904.3
Denmark	0.0	3.8	1.3	8.8	403.3	84.9
Estonia	45.0	0.0	0.0	0.0	79.1	66.9
Finland	402.9	12.9	1442.4	19.4	159.5	11325.2
France	1635.9	323.8	5614.1	370.4	3323.5	7253.0
Germany	9513.5	1082.2	16715.0	640.3	8157.2	16254.1
Greece	153.0	0.0	26.2	158.5	1073.3	102.5
Hungary	768.8	38.8	771.8	19.4	439.7	310.6
Ireland	0.0	0.0	0.0	19.4	507.3	9.0
Italy	2022.3	387.0	3385.2	131.9	5996.9	4938.8
Latvia	0.0	0.0	1.5	9.9	223.2	18.9
Lithuania	1126.2	0.0	0.0	0.0	239.9	70.7
Luxembourg	0.0	0.0	84.3	13.2	187.4	0.0
Malta	0.0	0.0	0.0	0.0	0.0	0.0
Netherlands	5142.7	0.0	3411.8	38.8	458.6	1376.9
Poland	3787.1	203.2	2883.0	19.4	3741.9	3169.0
Portugal	324.9	10.5	59.6	11.0	1136.2	3863.2

(continued)

Table A.1 (continued)

Country\Industry	Chemicals	Foundries	Iron and steel	Non-ferrous metals	Non-metallic minerals	Paper and pulp
Romania	813.5	0.0	1217.3	210.6	934.3	170.7
Slovakia	507.2	0.0	2219.9	162.4	703.3	1433.1
Slovenia	0.0	43.5	27.1	84.6	134.2	483.4
Spain	1682.0	144.2	2475.4	291.4	3195.7	3648.4
Sweden	485.1	71.8	1593.4	109.5	184.6	15084.7
UK	710.4	99.4	4596.7	58.5	1947.8	1091.2

Table A.2 Cumulative investment per country for EU27 + UK until 2030 (in EUR mn)

Country\Industry	Chemicals	Foundries	Iron and steel	Non-ferrous metals	Non-metallic minerals	Paper and pulp
Austria	72.4	13.4	257.3	4.3	297.6	516.6
Belgium	253.5	3.1	242.2	4.3	385.5	235.9
Bulgaria	27.2	1.6	12.7	4.3	151.4	34.9
Croatia	27.9	3.0	1.6	2.9	147.5	24.4
Cyprus	0.0	0.0	0.0	0.0	47.5	0.0
Czechia	21.0	18.1	185.8	4.3	239.1	96.3
Denmark	0.0	0.9	1.4	1.9	117.9	11.1
Estonia	3.0	0.0	0.0	0.0	23.5	7.1
Finland	35.7	2.9	128.6	4.3	82.2	1732.0
France	135.4	74.0	565.5	151.4	986.6	718.5
Germany	834.4	247.8	1551.5	231.2	2227.0	1831.9
Greece	10.3	0.0	10.0	64.6	318.6	12.3
Hungary	66.8	9.7	50.9	4.3	118.4	38.6
Ireland	0.0	0.0	0.0	4.3	150.6	2.0
Italy	178.3	94.1	518.3	29.2	1628.9	610.6
Latvia	0.0	0.0	1.4	2.2	66.3	2.3
Lithuania	76.0	0.0	0.0	0.0	66.9	8.5
Luxembourg	0.0	0.0	30.0	2.9	58.5	0.0
Malta	0.0	0.0	0.0	0.0	0.0	0.0
Netherlands	417.6	0.0	255.2	12.7	136.1	174.2
Poland	281.0	48.0	238.7	4.3	1025.0	369.4
Portugal	31.4	3.7	20.2	2.4	337.3	477.8
Romania	27.1	0.0	107.3	91.4	280.8	20.5
Slovakia	34.1	0.0	145.3	70.5	208.8	139.2
Slovenia	0.0	10.0	11.2	32.6	38.5	71.6
Spain	148.8	35.6	265.8	122.3	948.7	545.6
Sweden	24.5	16.1	148.1	43.4	84.9	2100.8
UK	87.0	23.0	354.0	20.7	578.2	219.4

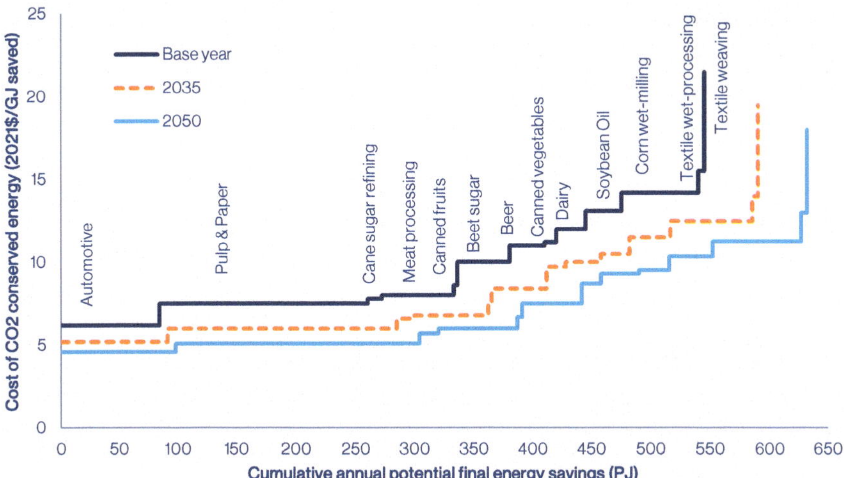

Fig. A.1 Energy saving potentials through heat pumps in US manufacturing

HPUS

Figure A.1[43] depicts the energy conservation cost curve, which displays the costs associated with energy savings resulting from the implementation of industrial heat pumps (IHP) applications across various US industrial processes. The chart shows the process-wide energy saving potential (in PJ) on the x-axis and the specific costs on the y-axis. In 2021, the technical potential energy savings resulting from IHP applications are estimated to be 545PJ per year, equivalent to approximately 4% of the current total final energy demand in US manufacturing. However, the chart indicates that IHP applications lead to additional costs in each process, with none of the industrial processes studied having energy-conservation costs falling below the horizontal axis (which would have represented cost savings). This implies that overall costs are not economical and require additional investment in IHPs for energy shift across all industrial sectors.

---

[43] Based on Figure 6.2 from Zuberi, Hasanbeigi and Morrow (2022). *Electrification through Industrial Heat Pump Applications in US Manufacturing*.

GCCA Roadmap

Actions to achieve Net-Zero concrete/cement (from GCCA Concrete Future Roadmap to Net Zero).[44]

Savings in process emissions
- Carbon capture and utilization/storage at cement plants (36%; 1370 Mt. CO_2)
- Efficiency in design and construction (22%; 840 Mt. CO_2):
  - Client brief to designers to enable optimization
  - Design optimization
  - Construction site efficiencies
  - Re-use and lifetime extension
- Efficiency in concrete production (11%, 430Mt CO_2):
  - Optimized mix design
  - Optimization of constituents
  - Quality control
- Savings in cement and binders (9%; 350 Mt. CO_2):
  - 80% of concrete's carbon footprint comes from cement
  - Alternatives to Portland cements (use industrial byproducts such as iron slag and coal fly ash)
- CO_2 sink: recarbonation (6%; 240 Mt. CO_2):
  - Natural uptake of CO_2 in concrete = a carbon sink

Savings in energy emissions
- Decarbonization of electricity (5%; 190Mt CO_2):
  - Decarbonization of electricity used both at cement plants and in concrete production
- Savings in clinker production (11% 410 Mt. CO_2):
  - Thermal efficiency
  - Savings from waste fuels ("alternative fuels")
  - Use of decarbonated raw materials
  - Use of hydrogen as fuel

---

[44] From Page 10 of Global Cement and Concrete Association (2022). *The GCCA 2050 Cement and Concrete Industry Roadmap for Net Zero Concrete.*

Emission-Reduction Potential by Industry

Fleiter et al. (2019)[45] quantify the emission reduction potentials in the EU for various industries under differing decarbonization scenarios. A partial implementation and combination of these measures can achieve an efficient decarbonization, which can be explored using the IndustryPLAN. It makes it possible to analyze partial implementations and combinations of measures and evaluate them versus their emission reductions and costs.

Cement potential: A more ambitious switch to low-carbon fuels such as biomass and maximum improvements in energy efficiency result in a 16% reduction in emissions by 2050 compared to 2015. A large-scale implementation of CCS can achieve a reduction of 81%. The use of synthetic methane and low-carbon cement types that replace Portland cement allows for a reduction of about 50% by 2050, but significant process emissions remain due to the continued use of low-carbon cement types that emit CO_2. A 62% reduction is possible by using biomass as the primary energy source and introducing material efficiency and recycling improvements in the construction industry, resulting in lower cement demand and production-related emissions. A switch to electricity combined with low-carbon cement types could result in a 45% reduction in emissions. A mix of measures that involve the use of electric furnaces for glass melting, low-carbon cement and material efficiency and recycling improvements could result in a 56% reduction. If that mix is complemented by using synthetic methane in the gas grid and CCS for any remaining conventional clinker and lime furnaces, it would result in an 86% reduction by 2050. Across all scenarios, the cement and lime production industries pose significant challenges to decarbonization, with low-carbon cement diffusion and material efficiency and recycling improvements in the construction industry being critical factors for reducing emissions.

Chemicals potential: The application of the best available technologies (BAT) in the chemical industry can achieve a 15% emission reduction by 2050 through energy efficiency improvements and increased use of biomass. Employing innovative strategies can lower emissions further. The use of carbon capture and storage (CCS) in all major processes can achieve a 90% reduction in emissions, including the negative emissions from biomass. The large-scale use of synthetic methane and a switch to hydrogen-based processes

---

[45] ICF Consulting Services Limited and Fraunhofer Institute for Systems and Innovation Research (ISI) (2019). *Industrial Innovation: Pathways to deep Decarbonisation of Industry.*

in ethylene and methanol production can reduce emissions by up to 77%. A comprehensive deployment of biomass and large improvements in material efficiency and circular economy, resulting in lower demand for energy-intensive products, can reduce emissions by 63%. Emission cuts of about 70% can be achieved by using hydrogen-based processes in combination with switching to the direct use of electricity for process heat generation. The challenges for decarbonizing the chemical industry are feedstocks, process emissions and the high share of natural gas, but hydrogen may play a key role in the industry's decarbonization.

Iron and steel potential: Implementing the best available technologies in the iron and steel industry could already result in a 51% reduction in emissions. This reduction is primarily due to the shift from oxygen steel to electric steel, accounting for 67% of total crude steel production by 2050. Replacing 88% of the oxygen steel production route with direct reduction based on hydrogen (DR H2 + EAF) reduces emissions by up to 88%. Replacing oxygen steel with electric arc furnace steel (EAF) is assumed to become available after 2030. A reduction of 69% can then be achieved with increasing material efficiency and the innovative use of electric steel for high-quality products, increasing the share of electric steel to 77% in total crude steel production by 2050. Further replacing oxygen steel with the alternative routes mentioned and additionally using synthetic methane to replace the remaining natural gas use achieves a reduction of 96%. Furthermore, the implementation of scrap-based steel production, taking advantage of the expected future increase in scrap availability, offers enormous potential for mitigation.

Pulp and paper potential: For the pulp and paper industry, the strict application of best available technology could achieve a 11% reduction by 2050 compared to 1990 by leveraging energy-efficiency improvements. By adopting carbon capture and storage (CCS) technology, the industry could achieve nearly 100% GHG emissions reduction, even if only about half of the paper mills combine bioenergy with CO_2 capture installations (BECCS). This high rate of reduction is attributed to the fact that the capture of CO_2 emissions from biomass results in negative emissions. Other measures include using electric steam boilers, replacing natural gas with synthetic methane, and phasing out coal-fired boilers and steam engines before their end-of-life after 2040. Given sufficiently high carbon prices, decarbonization through biomass and electricity usage could provide an additional business model for paper mills that have access to carbon storage sites and are equipped with CCS equipment: they have the potential to generate negative emissions through BECCS, which could compensate for process-related emissions in other industries.

6

Construction and Buildings
Green Blueprints

The EU's 2030 Climate Target Plan aims at slashing emissions in all sectors by a minimum of 55% by 2030 (compared to 1990), thus delivering the Fit-for-55 (Ff55) target that is part of the European Green Deal. Meeting the target would entail a 60% GHG emissions reduction for the buildings sector.

However, looking at the bloc's "fair share"[1] in achieving the Paris climate goals, emission reductions of at least 65% would be necessary in *all* sectors by 2030. Furthermore, the EU would need to achieve climate neutrality by 2040, 10 years earlier than planned under Ff55. Against this backdrop, the buildings sector could play a key role: Buildings account for 40% (residential: 26.3%, commercial: 13.7%) of the EU's total energy consumption and 36% of all energy-related GHG emissions.

The new REPowerEU Plan, introduced to accelerate energy sovereignty in the wake of the war in Ukraine, could lay the foundations for the buildings sector's green transition. Many of its measures address the diversification of energy supplies and propose energy savings through behavioral changes (e.g., less heating, taking shorter showers and the like). However, the key for long-term effects is the accelerated roll-out of renewable energy in buildings, using a mix of subsidies and mandates contained in various measures:

- Boosting national fiscal measures to encourage energy savings, such as reduced VAT rates on energy-efficient heating systems, appliances and products, and building insulation.

---

[1] The 'fair share' is calculated by the dividing the remaining carbon budget by the EU's share of the global population. Current targets imply that the EU would need to stay below 10% of the available global carbon budget to limit the temperature rise to 1.5 °C, given that the EU represents only 5% of the world's population.

L. Subran, M. Zimmer, *Investing in a Changing Climate*, Professional Practice in Governance and Public Organizations, https://doi.org/10.1007/978-3-031-47172-8_6

- Increasing the binding EU Energy Efficiency Target from 9% to 11.7% in 2030 vs. 2020.
- Implementing a solar rooftop initiative, with a phased-in legal obligation to install solar panels on new public and commercial buildings and new residential buildings.
- Doubling the rate of heat pump deployment, with the aim of at least ten million additional heat pumps by 2027 and 30 million additional heat pumps by 2030 compared to 2020, and introducing measures to integrate geothermal and solar thermal energy in modernized district and communal heating systems.
- Extending buildings' Minimum Energy Performance Standards to all buildings, and particularly strengthening the national energy requirements for new buildings (with the aim of not only having zero-emission buildings but also energy-plus buildings that produce more renewable energy than they consume).
- Tightening national heating system requirements for existing buildings.
- Moving forward the end of member states' subsidies for fossil fuel-based boilers from 2027 to 2025, as a first step towards a complete ban on such boilers in both existing and new buildings shortly thereafter.

These measures focus on the two crucial approaches for achieving GHG emission-reductions in the buildings sector: (1) improving energy efficiency, and (2) switching to greener fuels.

Renovation is another factor which will play a key role in improving energy efficiency. Heating and cooling account for 80% of the energy consumed in residential buildings, largely because of the continued reliance on fossil fuels and outdated technologies and appliances. Renewable heating and cooling made up only 22.9% of buildings' gross final energy consumption in 2021. Hence, one of the key focus areas is to slash the energy demand for heating and cooling by renovating the building envelope (e.g., insulation and windows), which is particularly important for older buildings that were constructed under lower energy-performance standards. In addition, the equipment fitted and used in buildings also matters: switching to electric heating systems, adopting renewable solutions such as solar systems, using standardized and labelled energy-efficient equipment and smart-home technologies (e.g., for controlling lighting and heating) can all play an important role in keeping energy demand in check.

If the EU is to achieve its original 2030 emission targets, significant final energy savings are necessary (Fig. 6.1). The greatest savings would be in the residential sector, with 24% lower total energy consumption in buildings, and 31% savings in heating and cooling.

Fig. 6.1 Evolution of energy consumption in buildings to 2030 (compared to 2005). Sources: EU Commission 2030 Climate Target Plan, Allianz Research

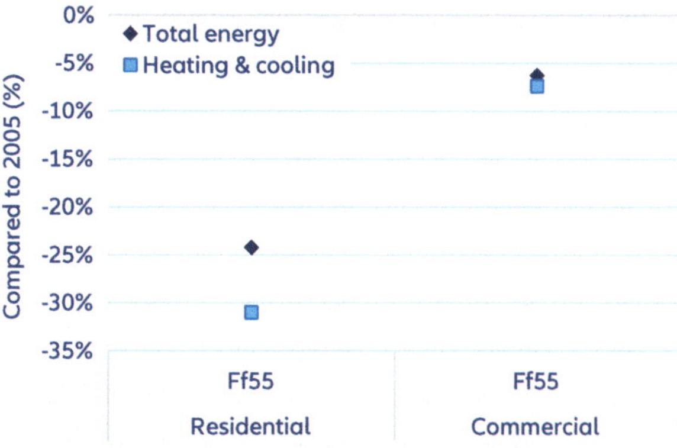

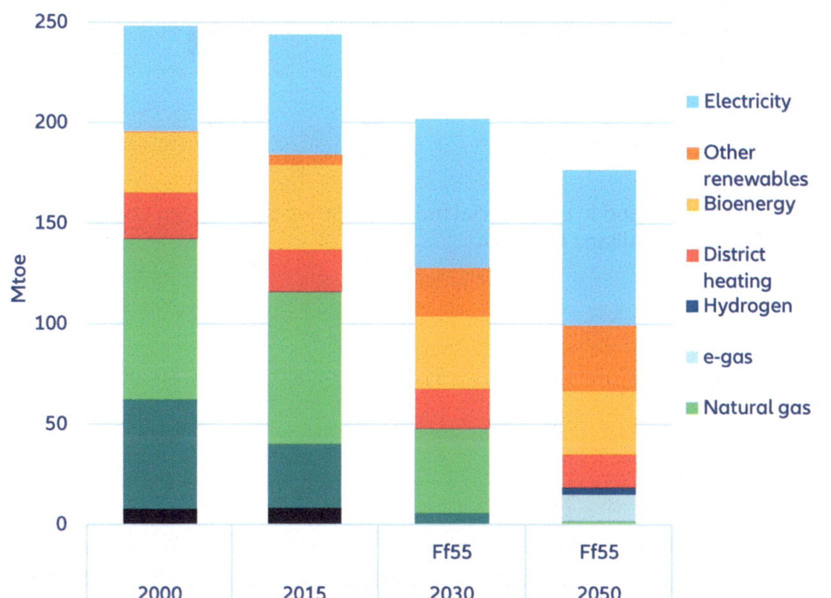

Fig. 6.2 Energy demand in residential buildings. Sources: European Commission 2030 Climate Target Plan, Allianz Research

Greening up the fuel mix of buildings will entail rapid growth in electricity consumption as fossil fuels decline (especially natural gas). For residential buildings, the electricity share is expected to increase to 36% by 2030 (+15 pp), topping 45% by 2050 (Fig. 6.2). In commercial buildings, the share of electricity will reach 50% by 2030 (+5 pp) and 60% by 2050 (Fig. 6.3).

The expected rise in electricity-based heating, coupled with overall declining energy demand, will drive down conventional heating and, with it, the consumption of other fuels (particularly fossil fuels) (Fig. 6.4).

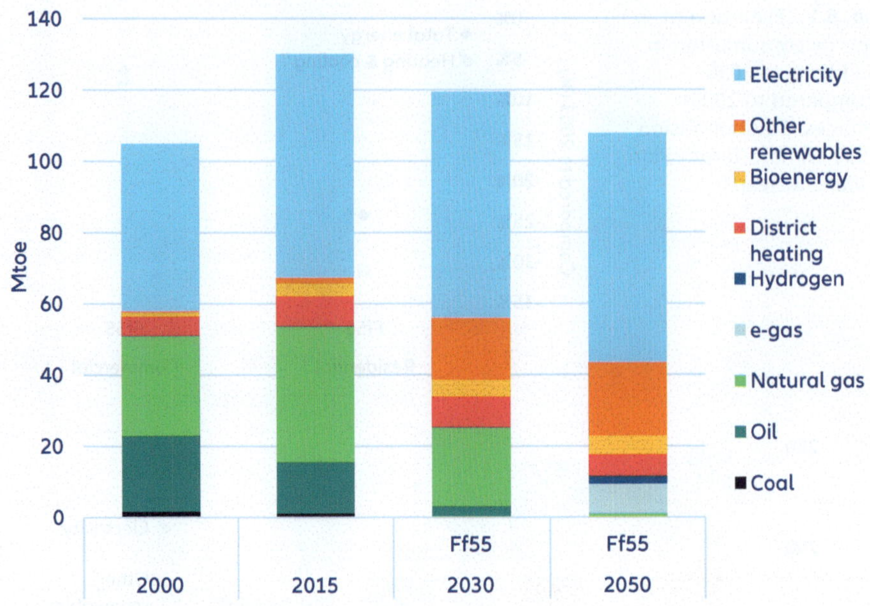

Fig. 6.3 Energy demand in commercial buildings. Sources: European Commission 2030 Climate Target Plan, Allianz Research

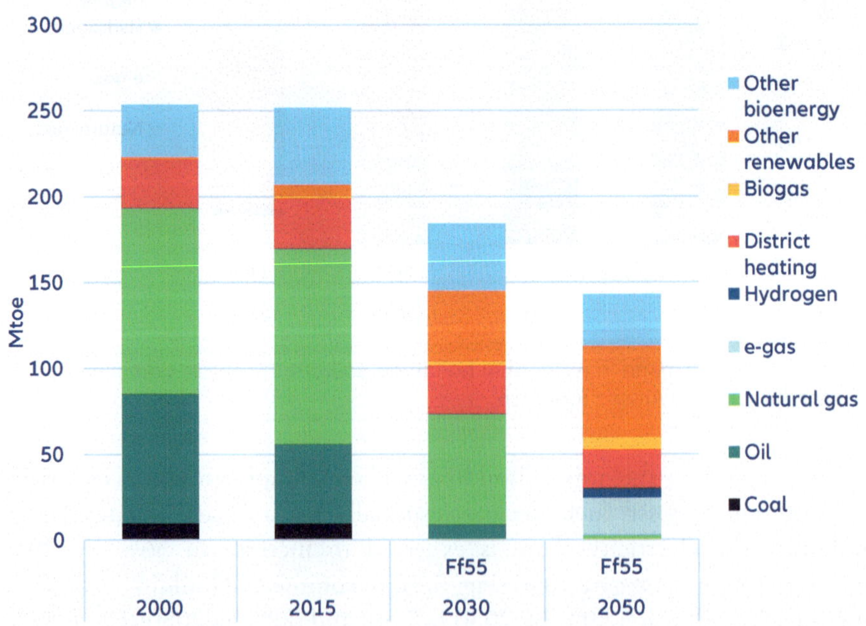

Fig. 6.4 Non-electricity energy consumption in residential and commercial buildings. Sources: European Commission 2030 Climate Target Plan, Allianz Research

The required energy savings and the switch to renewables can be characterized by energy-reduction and decarbonization pathways. Figure 6.5, 6.6, 6.7, and 6.8 depicts the commonly used CRREM (Carbon Risk Real Estate Monitor) energy-intensity and emission-intensity pathways for office buildings, and compares them to the pathways from the One Earth Climate Model (OECM) for commercial as well as residential buildings. The pathways shown claim to be consistent with keeping global warming below 1.5 °C. While CRREM features pathways by country, the OECM only shows a joint OECD Europe pathway. However, the evolution regarding energy intensities is reasonably similar. Interestingly, the pathway for residential buildings in OECM starts on a much lower intensity level and does not need to decline as steeply to converge to a similar absolute level in 2050 as the commercial energy intensity one.

The picture looks different for the emission intensity, though. While OECM and CRREM seem to start on similar average levels in 2020, the OECM

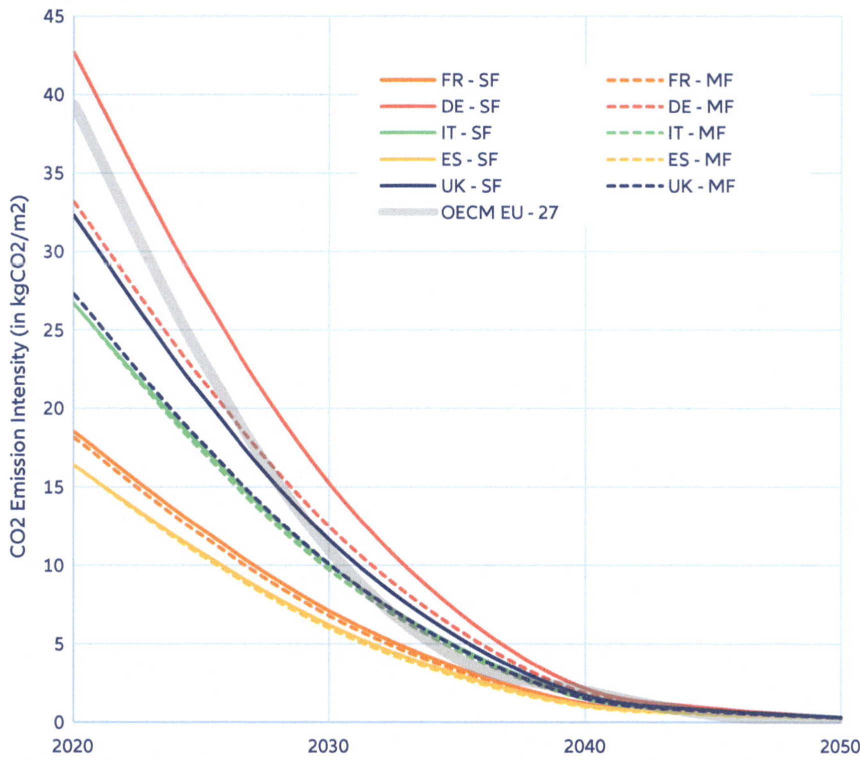

Fig. 6.5 Carbon intensity of single- (SF) and multi- (MF) family residences. Sources: CRREM, OECM, Allianz Research

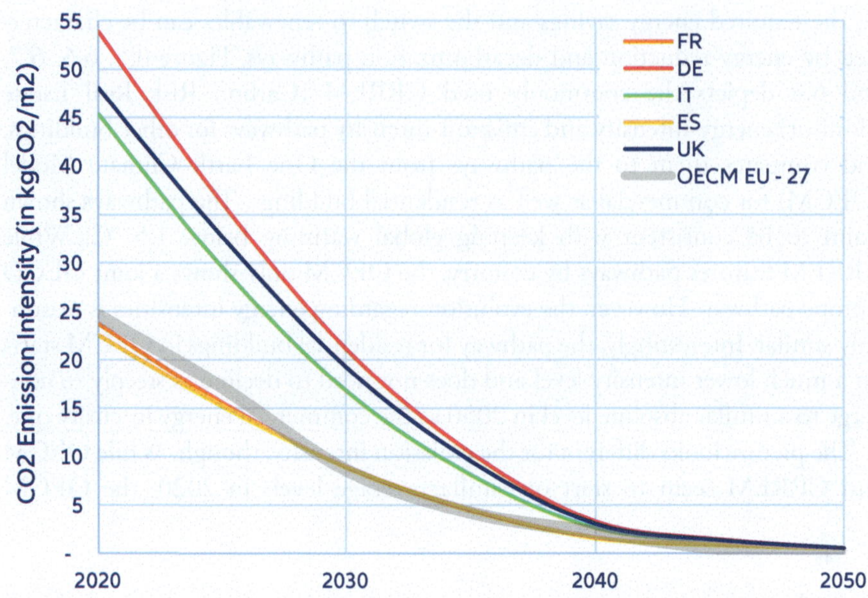

Fig. 6.6 Carbon intensity of commercial buildings. Sources: CRREM, OECM, Allianz Research

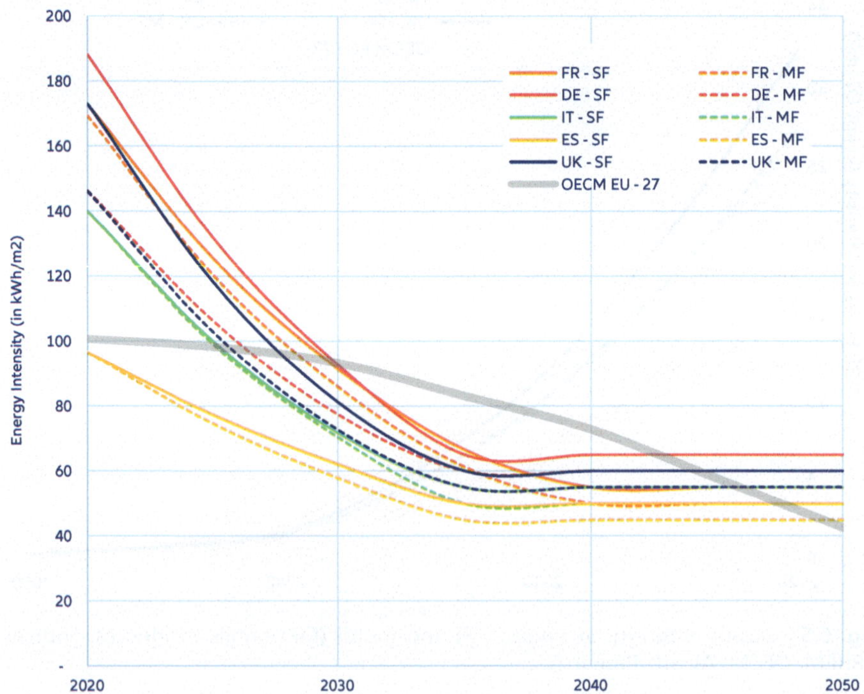

Fig. 6.7 Energy intensity of single- (SF) and multi- (MF) family residences. Sources: CRREM, OECM, Allianz Research

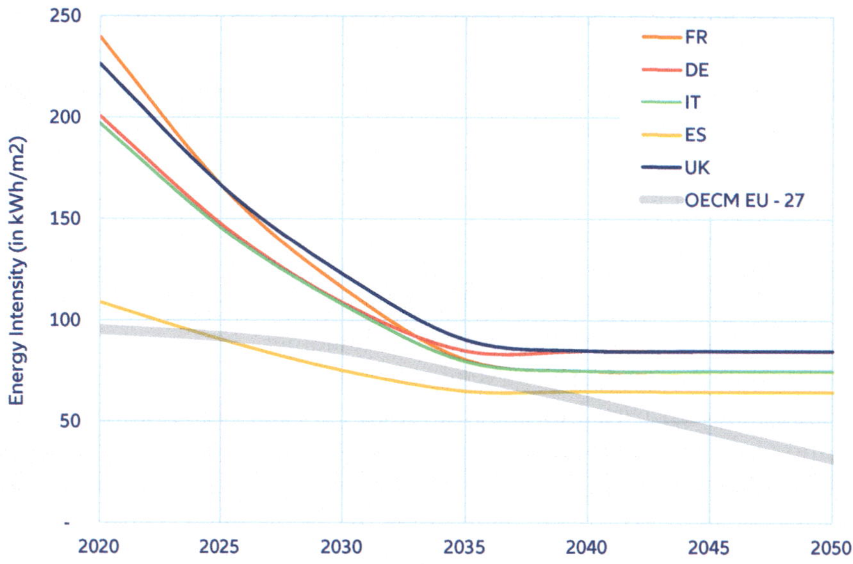

Fig. 6.8 Energy intensity of commercial buildings. Sources: CRREM, OECM, Allianz Research

emissions decline much faster and also converge to zero in 2050, while the CRREM emission-intensity pathways converge to just below $3kgCO_2e/m^2/yr$. This raises a question regarding the reason for this difference. The simple answer is that both frameworks operate under different assumptions, with the CRREM pathways relying on rather outdated climate pathways. These shortcomings underline the potential need for ramping up decarbonization ambitions by more than what is suggested by the CRREM projections.

Build Back Better

According to the EU Building Stock Observatory, the total building stock, comprising dwelling units and non-residential buildings, amounted to about 260 m units in 2016–17. Around 85% of these, roughly 220 m units, were constructed before the beginning of the twenty-first century, with up to half of the building stock dating back to the pre-1960s. Up to 95% of the existing buildings are expected to still be around in 2050.[2]

The Energy Performance of Buildings Directive (EPBD) requires all new buildings from 2021 onwards (public buildings from 2019) to be 'nearly

---

[2] European Commission (COM(2020) 662 final). *A Renovation Wave for Europe - greening our buildings, creating jobs, improving lives. European Commission.*

zero-energy buildings' (NZEB). However, every year, only around 1% of the existing building stock is newly constructed. Therefore, the only way to achieve the Paris climate target of 1.5 °C goes through renovating the older, inefficient and energy-intensive buildings. The energy performance certificates (EPCs) for buildings, introduced by the EU in 2002 (ranging from A = most efficient to G = least efficient), is a tool to identify the most "attractive" renovation targets: across the EU, 35% of buildings have an EPC rating between D and G. A proposed revision to the EPC aims to define the G rating as the worst-performing 15% of buildings, requiring their upgrade until 2027 for non-residential and 2030 for residential buildings.

This would be a big step forward. Currently, around 16% of the EU's buildings undergo partial to full renovation per year. However, little to no attention is being paid to assessing and improving the energy performance of those buildings: while three-quarters of renovations were energy-related, more than half of so-called 'energy renovations' are below the 3% energy-saving threshold to be considered as actual energy-saving. Another one-third consists of 'light energy renovations,' with savings between 3% and 30% per annum. Less than 10% have resulted in 'medium energy savings' of 30%–60%, and only around 2% in 'deep energy savings' exceeding 60%. Consequently, weighted by effective energy saving, the annual energy renovation rate was estimated at close to 1% for residential buildings and close to 0.5% for non-residential buildings, according to the European Commission[3] (Fig. 6.9). These rates need to at least double to achieve the EU's climate targets.

Limits to Energy Efficiency

As a major energy consumer, improving the sector's energy efficiency should be a no-brainer. It is important to note, however, that implementing such measures does not always guarantee energy conservation. This is due to the so-called *rebound effect*, when beneficial energy savings resulting from improved energy efficiency are counteracted by a rise in energy consumption of similar magnitude. In the context of the building sector this means that if the costs of energy do not vary greatly, people often tend to use more energy when equipped with energy-efficient equipment, since the energy bill remains the same even at a higher consumption level.

The magnitude of the rebound effect (R) can be calculated as the percentage increase in energy consumption for every percentage increase in energy efficiency for a specified period.

---

[3] European Commission (2019). *Comprehensive study of building energy renovation activities and the uptake of nearly zero-energy buildings in the EU – Final Report.*

- If R > 100% = > backfire (energy efficiency gains lead to an increase in energy consumption)
- If R = 100% = > full rebound
- If 0% < R < 100% = > partial rebound
- If R = 0% = > no rebound (actual energy savings match the predictions)
- If R is negative = > super-conservation (energy-efficiency measures result in a larger-than-expected drop in energy demand)

The existence of rebound effects (R > 0%) could keep the EU from reaching its emission-reductions targets on time. A study on rebound effects for energy-efficiency measures in the EU's residential sector notes big differences between countries. In the first group (Bulgaria, Czechia, Estonia, Hungary, Italy, Romania, Slovenia, and Spain), a rebound of 100% or more is observed, which means that the measures usually resulted in a full rebound or even backfired. In the second group (Austria, Denmark, Croatia, Germany, Greece, Latvia, Lithuania, and Poland), a partial rebound effect is observed. Only four countries (Belgium, Finland, Ireland and Luxembourg) showed almost no rebound effects.[4]

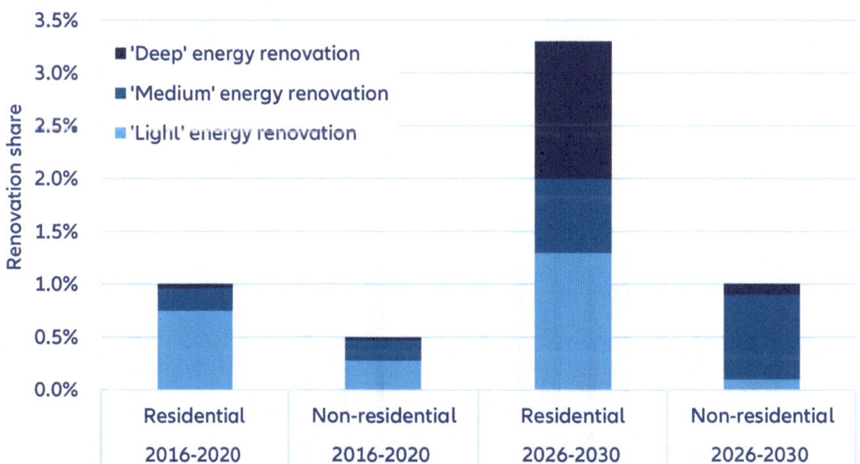

Fig. 6.9 Development of weighted energy renovation rates (Type 1*) in buildings, by sector. Source: EU Commission [Underlying data has been sourced from: European Commission (2019). Comprehensive study of building energy renovation activities and the uptake of nearly zero-energy buildings in the EU – Final Report.], Allianz Research, * Only renovation of the building shell (e.g., insulation)

---

[4] Baležentis et al. (2021). *Exploring the limits for increasing energy efficiency in the residential sector of the European Union: Insights from the rebound effect.* Energy Policy, 149, 112,063. https://doi.org/10.1016/j.enpol.2020.112063

Investment Needs to Close the Gap

Renovations do not come cheap. An estimated 30 m building units are in the worst-performing Category G energy label and require renovations under the EPBD. The EU budget has allocated up to EUR150bn to support the upgrading of these buildings until 2030. Given that the EU's construction industry employs 18 m people and contributes about 9% to total GDP, such a wave of renovations could potentially create 160,000 new jobs in the construction ecosystem.

As Fig. 6.10 shows, the actual investment for a deep energy renovation (the kind that results in more than 60% energy saving) varies widely between EU countries. A large-scale analysis of actual renovation projects placed Spain among the cheapest (around EUR50 per m^2 for residential projects) and Sweden among the most expensive (EUR450 per m^2 for non-residential projects). Except for the UK, non-residential projects were more expensive even when normalized to square-meter values.

But how much bang do you get for the buck? Figure 6.11 shows how many euros need to be spent to save 1 kWh of energy or to mitigate 1 kg of CO_2,

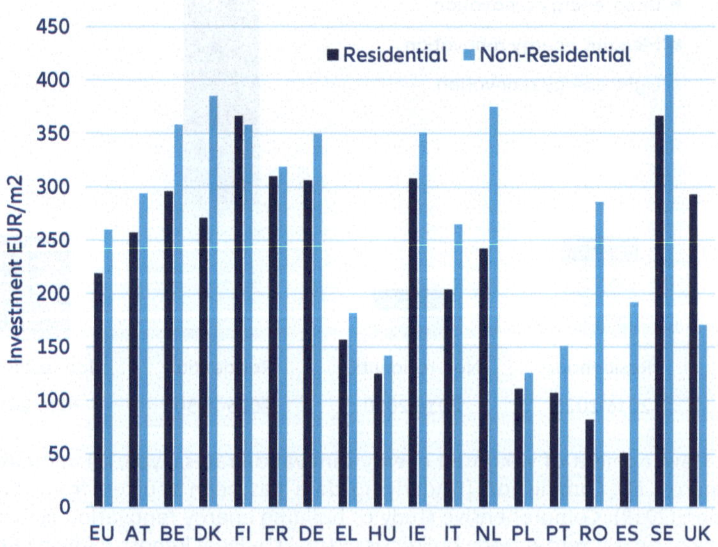

Fig. 6.10 Investment needs for 'deep' energy-related renovation,* by country. Sources: EU Commission, Allianz Research, own calculations [Own calculations based on: European Commission (2019). Comprehensive study of building energy renovation activities and the uptake of nearly zero-energy buildings in the EU – Final Report. Data and calculation sheet can be provided upon request]. * More than 60% energy saving

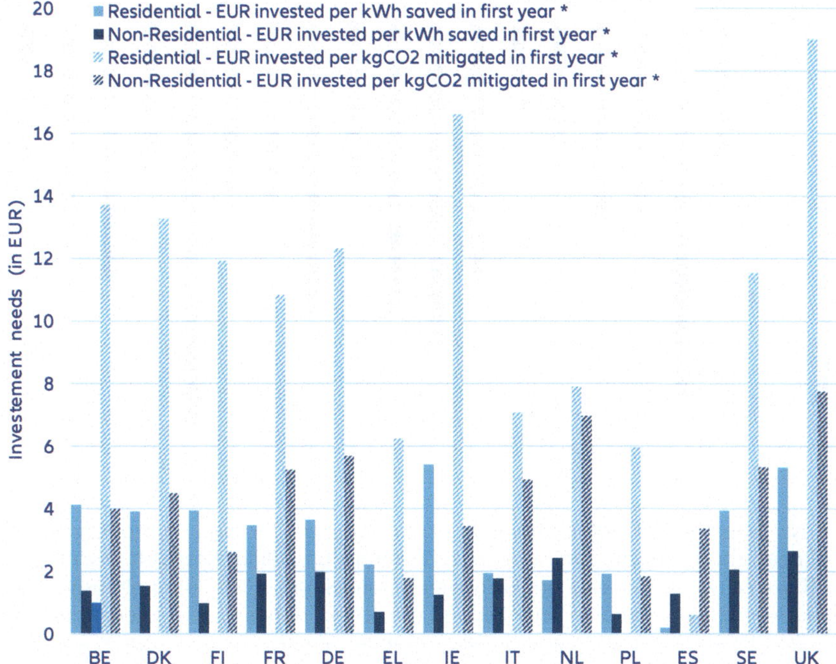

Fig. 6.11 Investments in deep energy renovations required for energy savings and emissions mitigation*. Sources: EU Commission, Allianz Research, own calculations [Own calculations based on: European Commission (2019) Comprehensive study of building energy renovation activities and the uptake of nearly zero-energy buildings in the EU – Final Report. Data and calculation sheet can be provided upon request]. * The investment needs are relative to the energy and emission savings in the first year after the renovation and thus do not reflect the mitigation costs per unit of energy or emissions over the lifetime of the investment. These are much lower as the investments produce annual savings over decades

respectively. Naturally, these figures closely correlate with the investment costs per square meter, given the definition of the deep energy-renovation category (more than 60% energy saving) and the fact that the actual observed energy savings and emissions mitigation in that category do not vary dramatically from country to country. Energy savings of 1 kWh per year were mostly realized at investments of between EUR1-EUR5. Reducing CO_2 emissions by 1 kg per year was almost three times as expensive in all countries.

Energy savings are per year and permanent (within the life cycle of the investment) and thus cumulative over time. Given the different investment needs for energy savings—and the varying consumption-weighted energy costs for households, ranging from 5 cents per kWh in Hungary to 26 cents per kWh in Sweden—the time to recoup the investment purely through

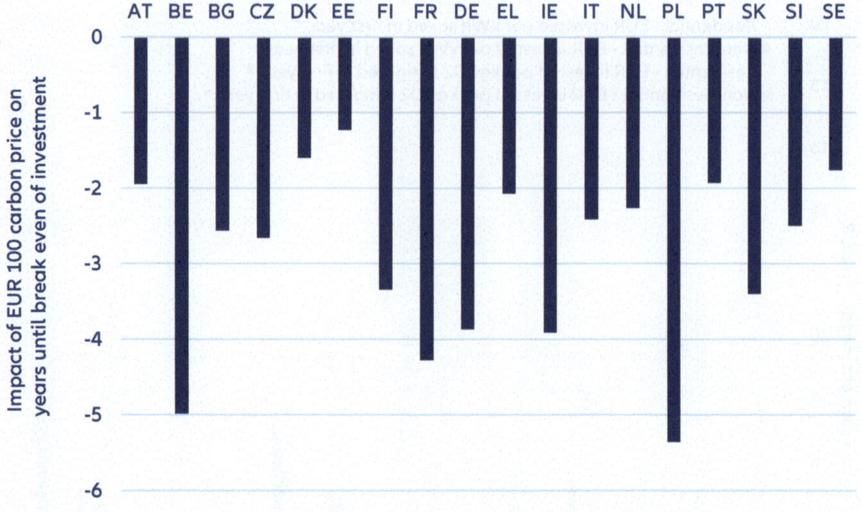

Fig. 6.12 Effect of EUR 100 additional carbon price pass-through on breakeven horizon of deep energy renovation. Sources: EU Commission, Allianz Research, own calculations [Own calculations based on: European Commission (2019). Comprehensive study of building energy renovation activities and the uptake of nearly zero-energy buildings in the EU – Final Report. Data and calculation sheet can be provided upon request]

energy savings differs widely across the EU, with an average of around 16 years (this excludes subsidies for energy renovations and the most recent rise in construction costs and energy prices). Figure 6.12 shows the effect on the time-to-breakeven of deep energy investments if an additional carbon price of EUR100 would be passed through energy bills to consumers. For most countries, this would reduce the number of years to recover investments by 2 to 4 years, with the average in our sample lying at 3.3 years.

Figure 6.13 shows the annual energy-renovation investment needed to achieve the EU goal of a 2% energy-renovation rate. This estimate uses the aspired investment mix between light, medium and deep renovation (Fig. 6.9 above). About 82% of investments are supposed to flow into residential energy renovation. Investment needs vary between 0.2% and 0.8% of GDP, with an average of 0.5%. In total, this amounts to EUR82bn per year in the EU, or EUR95bn if the UK is included.

However, as discussed above, the EU's climate ambition falls below its fair contribution to limiting global warming to 1.5 °C. Furthermore, the Russian invasion of Ukraine made the energy transition even more urgent as the only viable path to energy sovereignty. In this context, the ambition should be upped to a 3%-per-year energy renovation target for the building stock.

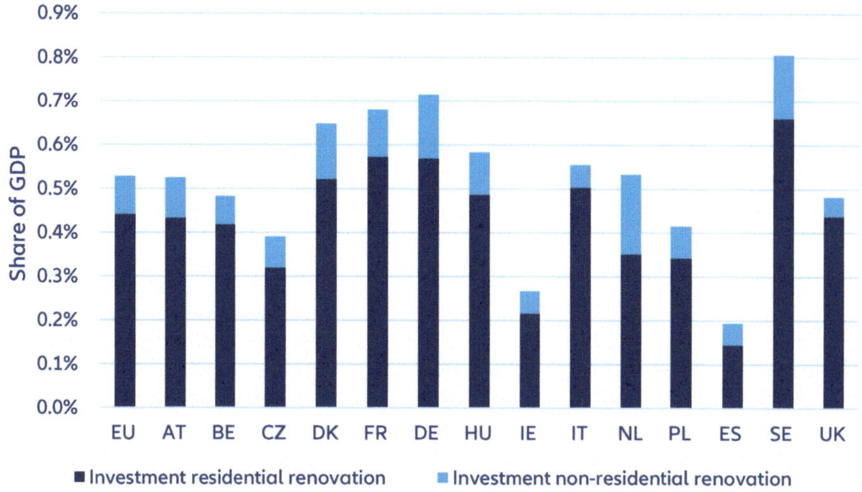

Fig. 6.13 Annual energy-renovation investments at 2% renovation rate and the planned EU renovation mix. Sources: EU Commission, Allianz Research, own calculations [Based on European Commission Report (see "Own calculations based on: European Commission (2019). Comprehensive study of building energy renovation activities and the uptake of nearly zero-energy buildings in the EU – Final Report. Data and calculation sheet can be provided upon request")]

Table 6.1 translates this "ideal" ambition into concrete action by comparing the current EU targets against the progress made based on the three key levers "buildings renovated", "heat pumps installed" and "solar rooftops installed", as well as with the progress that should be reached for climate and energy sovereignty (i.e., a 3% energy renovation rate). Our analysis confirms previous findings and suggest that[5]:

- 5 m buildings per year need to undergo substantial energy renovation.
- 5 m heat pumps need to be installed, mainly to replace fossil-fueled heating systems.
- 5 m solar rooftops need to be installed.

Although bold compared to current targets, the new ambition is still feasible for a simple reason: the EU's overall goals do not seem particularly impressive when compared with current market developments and policies announced by individual EU member states. In particular, the deployment of heat pumps

---

[5] See for instance CAN Europe (2022). *Repowering for the people – Flagship actions the Commission's plan 'REPowerEU' should feature in the current fossil fuel and energy prices crisis.*

Table 6.1 Concrete measures by different ambition levels

All values per year	EU 2021 pre Ff55	EU Ff55 pre-war	EU Ff55 post-war	3% renovation rate	Remarks
Buildings renovated	1.75 m	3.2 m	3.5 m	over 5 m	Renovations with substantial energy savings. According to the COM(2020) 662 final communication by the European Commission, a 2% renovation rate equates to 35 million buildings in 10 years.
Heat pumps installed	2.0 m	1.0 m	2.0 m	over 5 m	EU less ambitious than expected actual market development.
Solar rooftop installations	1.4 m	1.2 m	1.4 m	over 5 m	At an assumed 10kWp per installation. Low ambition by the EU as Germany alone plans to ramp up installation to over 1 m solar rooftops per year. Market peaked already at 1.4 m in 2011 and declined to below 0.3 m in 2014–2016.

is seeing large growth rates. In addition, the German government's aim to add 11GW of solar rooftop capacity per year would on its own satisfy most of the EU's capacity expansion target.

Energy renovations are typically triggered by other renovations, unexpected repairs, regular maintenance, and inspections. Other important factors include budget availability or health issues that would require a modification of the premises. In the past, the most relevant aspects of residential energy renovation were not the energy saved, but the associated cost savings as well as making the home more comfortable and healthier to live in. Thus, even within the current renovation budgets and capacities, the revised targets are achievable if the focus of renovation activities is placed more firmly on energy performance improvements.

Concretely, the 3% renovation rate proposed, which implies over 5 m energy-renovated units per year, suggests an investment gap of EUR47bn per year in the EU-27 plus UK (i.e., total investment needs of EUR142bn per year. See also Fig. 6.13 for a 2% renovation rate). However, the EU's residential sector currently spends about EUR200bn per year on energy-related renovations, mainly on rather inefficient 'light' ones, and EUR300bn on non-energy-related renovations. Another EUR200bn is being invested in renovating non-residential buildings.[6] In this context, closing the investment gap should be possible if improving energy performance becomes the first principle when undertaking building renovations.

Roadblocks to Energy Renovation

Even if the overall investment sums are not exorbitant, there are other roadblocks to the quick scale-up of energy-related renovation activity. There is a good deal of uncertainty regarding what to expect from installers, with a high share of laypeople (tenants, homeowners) taking responsibility for quality control on the one hand, and, on the other, half of all installers across Europe reporting energy-efficiency measures as being too complicated to install. It does not help that incentives for renovations (be they economic or related to regulations or living quality) rarely take account of the demand-side split into tenants, owner-occupiers and landlords on the one side, and on the other the supply-side division of labor between architects, main contractors and installers. As it turns out, consumers, architects, contractors and installers across the board view financial and administrative barriers as being the main roadblocks for undertaking effective energy performance improvements on buildings.

It is striking that a vast majority of consumers use their own capital to finance renovation works, suggesting that consumers refrain from undertaking energy-related renovations unless they have sufficient own capital available. For commercial clients, financing and savings are even stronger motivations than for residential clients.

And with good reason. Construction costs have been steadily rising over the years, with a notable 15% increase between 2010 and 2019. Taking the construction costs for new residences as representative for renovations, Fig. 6.14 shows the evolution of the construction cost index (CCI). However, the 2010–2019 increase can hardly be compared to the price rally seen since the third quarter of 2021. Due to the combined effects of the massive disruption in global supply chains and the energy price explosion caused by the Russian invasion of Ukraine, construction costs are now (Q1–2023) 45% higher than in 2010. The Austrian CCI curve, close to the EU average between 2010 and 2019, currently indicates an expected 65% price increase for Q1–2023 vs. Q1–2010. On the positive side, energy-efficient buildings will insulate residents somewhat from rising energy prices. Furthermore, higher energy prices not only increase construction prices, but also lead to a quicker return on renovation investments. This means that the market for decarbonizing buildings can only become more attractive.

---

[6] We base our calculations on the specific energy efficiency-related investment needs derived from 'deep energy renovations'. Other studies include the non-energy efficiency-related investment part of the dominant 'medium-energy renovations' in the investment volume and feature higher investment figures, e.g., EUR243 billion in BPIE (2022). *A Guidebook to European Buildings Efficiency: Key regulatory and policy developments - Report on the evolution of the European regulatory framework for buildings efficiency*, or EUR275 billion in European Commission (2022). *Revised Construction Products Regulation – Factsheet.*

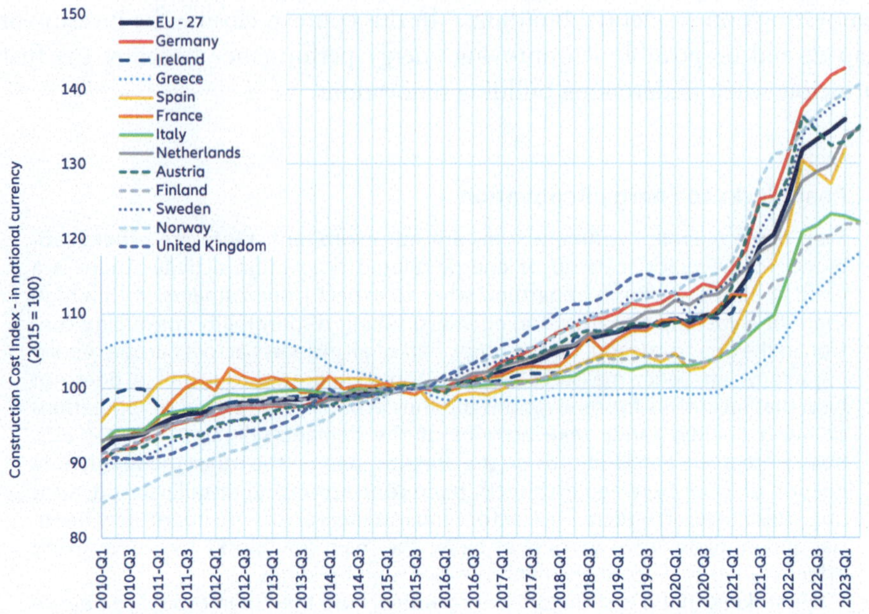

Fig. 6.14 Construction cost index (CCI)—new residential buildings. Sources: Eurostat (STS_COPI_Q), Allianz Research

Going Full Cycle

The sector's ultimate goal should be to transform all buildings into carbon-neutral units, not only throughout their operational life but over their entire life cycle (encompassing the construction phase, production and transportation of the emission-intensive materials used to build them, as well as the renovation and demolition phases). A briefing by the European Environment Agency (EEA, see Appendix) suggests that by taking measures such as reducing the amount of cement, concrete, steel and aluminum used in the construction process, buildings could cut the materials-related GHG emissions by 61% over a building's life cycle by 2050. During the design phase, this could be accomplished by reducing overspecification of concrete in building plans (−12%); in the production phase, another cutback could be achieved using innovative and alternative cement types (−16%); in the demolition and waste management phase, re-use of structural steel would also make a sizable contribution (−15%). These three actions focus directly on making the use of steel, cement and concrete more efficient, as these constitute the most emission-intensive materials used in the buildings sector. As these emissions appear in the supply-chain of the construction sector, they are not reflected in the sectoral emission pathways shown in Figs. 6.5, 6.6, 6.7, and 6.8 which focus on the direct energy use within the buildings. Still, they form part of investors' life-cycle assessments and count against their climate targets.

The European Commission does promote sustainable building materials and solutions. It has a plan to focus on the sustainability performance of construction products under the revised Construction Products Regulation (CPR), which is currently in the adoption process. The regulation can be seen as roadmap to 2050 for reducing whole-life-cycle carbon emissions. Materials' recovery targets for construction and demolition waste (CDW) in EU legislation will also be reviewed by the European Commission by the end of 2024.

In addition to construction materials, a more intensive focus will be placed on insulation materials used in renovations. While such materials improve the energy efficiency of buildings, any potential trade-offs, such as the emissions generated in the production of insulating materials, must also be factored in for a holistic picture of life-cycle emissions.

Intriguingly, the carbon footprint of a building need not stop at zero on the lower end: Recent research has highlighted the role of buildings as potential carbon sinks.[7] As it turns out, buildings could actually have negative net-emissions—if made of wood. A five-story residential building structured in laminated timber can store up to 180 kg of carbon per square meter, three times more than in the above-ground biomass of high-carbon-density natural forests. The study also suggests that close to 700 m tons of carbon could be stored per year if 90% of new buildings worldwide were made of wood.

While it is true that the carbon stored in buildings over 30 years would add up to less than one-tenth of the overall amount of carbon stored above ground in forests worldwide, and 700 m tons seems small compared to the current global emissions of roughly 10,500 m tons of carbon (equaling almost 38,500 m tons of carbon dioxide) per year, every small effort counts.[8] Home green home indeed.

---

[7] See, for example, Churkina, G., Organschi, A., Reyer, C.P.O. et al. (2020). *Buildings as a global carbon sink*. Nature Sustainability 3, 269–276. https://doi.org/10.1038/s41893-019-0462-4

[8] See also: IEA. *Energy Systems – Buildings*.

Conclusions

Buildings' total energy consumption and energy-related GHG emissions make them a prime target for decarbonization. But, given that 85% of the building stock is energetically inefficient and 95% of today's building stock is expected to be still in use in 2050, the task is neither easy nor quick to achieve.

New buildings are already subject to stringent mandates, so the focus falls in particular on renovating the existing stock, aiming for improving not only energy efficiency but also a switch to greener fuels or to more efficient technologies. Heat pumps are among such technologies. Energy performance certificates could help to spur change.

Most energy renovation, however, falls currently under the "light" category—anywhere from below the 3% energy-saving threshold to be considered as actual energy-saving to a still insufficient 30%—since the cost of deep renovation can be high, and varies greatly among EU countries. In response, the EU is allocating EUR 150bn for upgrading the worst-performing building units until 2030.

Some education and training will also be needed. First, on the demand side, by making users aware of the "rebound effect", which results from dwellers consuming more energy when efficiency improves. Second, on the supply side, among all those involved in carrying out the renovations: architects, contractors and installers.

Circularity must also play a role, which can be achieved only when the entire lifecycle of a building is taken into account, from its design to demolition and recycling.

If done well, buildings can also be turned from carbon emitters to carbon sinks, since the materials they are made of, if chosen properly, can act as carbon storage. Wood would be the material of choice for this purpose.

Appendix

Cutting greenhouse gas emissions through circular economy actions in the buildings sector

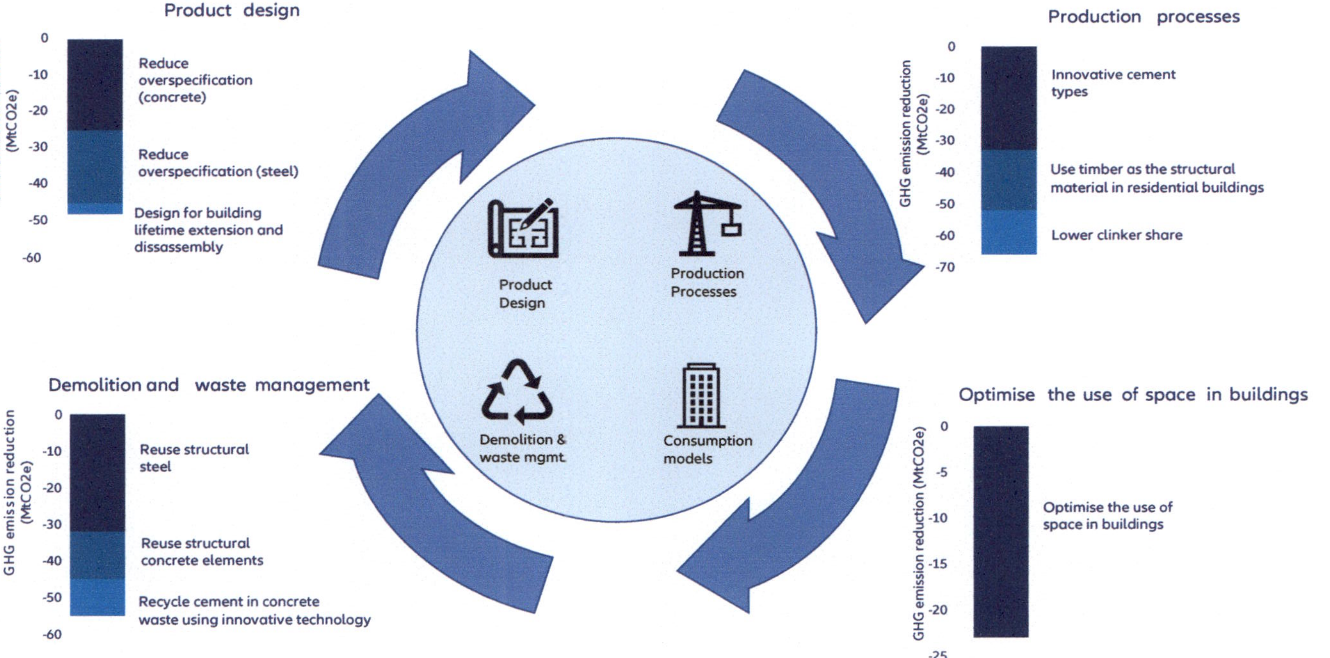

Sources: EEA, Ramboll,[9] Allianz Research

[9] Ramboll, Ecologic Institute and Fraunhofer ISI (2020). *The Decarbonisation Benefits of Sectoral Circular Economy Actions.*

7

Forestry, Agriculture, Food Chain, and Land Use
Greener Pastures

Agriculture is arguably the foundation of our civilization. Throughout history, civilizations have risen (and fallen) based on their agricultural productivity and land management: a civilization that fails to feed itself is usually doomed. Furthermore, growth in agriculture has historically tended to be followed by wider economic development. Nowadays, close to 38% of the 13 billion hectares of the world's land area is devoted to agriculture.[1] A 2009 report by the United Nation's Food and Agriculture Organisation suggested that by 2050 agricultural production would have to rise by 70% to meet projected demand. But given that most land suitable for farming is already farmed, this growth must come from higher yields.

Across the EU-27, agriculture dominates land use, covering an average of around 39% of the EU's total area in 2018, followed by around 35% for forestry.[2] The agriculture and forestry sectors also supported around 10 m jobs in 2019 and accounted for around 1.5% of the EU's GDP.

As it turns out, both farming and forestry are the sectors most under threat from the impacts of climate change, given their sensitivity to weather patterns. Higher temperatures and carbon dioxide concentrations, uncertain rain patterns and the frequency of extreme weather events are all on the rise. Volatility is preordained.

Across the EU, the following effects are expected:

---

[1] FAO (2021). *Statistical Yearbook 2021.*

L. Subran, M. Zimmer, *Investing in a Changing Climate*, Professional Practice in Governance and Public Organizations, https://doi.org/10.1007/978-3-031-47172-8_7

Mediterranean region—Portugal, Spain, Italy, Greece, Croatia

- Increase in heat extremes, drought, water demand, and biodiversity loss
- Decrease in precipitation, crop yields

Continental region—Germany, eastern France, Czech Republic, Poland

- Increase in heat extremes, river floods
- Decrease in summer precipitation

Atlantic region—Western France, Belgium, Netherlands, Northern Germany, UK

- Increase in heavy precipitation events, risk of river and coastal flooding, damage from winter storms

Boreal region—Sweden

- Increase in heavy precipitation events, damage from winter storms, lower crop yields

Mountain regions—Norway, European Alps (Switzerland, Italy, Austria), northern Hungary

- Larger temperature increases than EU average
- Migration of plant and animal species to higher altitudes
- Increase in hail risk, risk from rockfalls and landslides

EU agriculture contributes significantly to the global food supply, thanks to high productivity. EU farmers are estimated to produce one-eighth of the global cereals output, two-thirds of global wine and three-quarters of the world's olive oil. EU agricultural trade was close to EUR350bn in 2021 (with exports almost reaching EUR200bn) and accounted for 8% of total EU international trade.[3] But as climate conditions change, productivity will change, too, albeit mostly in a negative way. On a global scale, a recent study[4] observed that based on the current climate change prospects, corn yields could decline by as much as 24%

---

[2] IMF (October 2021). *World Economic Outlook database.*

[3] See Eurostat (Update 03/2023). *Extra-EU trade in agricultural goods* and Eurostat (Update 07/2023). *International trade of EU, the euro area and the Member States by SITC product group.*

[4] Jägermeyr, J., Müller, C., Ruane, A.C. et al. (2021). *Climate impacts on global agriculture emerge earlier in new generation of climate and crop models.* Nat Food 2, 873–885. https://doi.org/10.1038/s43016–021-00400-y

by 2030. The yield projections for soybeans and rice suggest that they will also decline in some regions. Meanwhile, wheat could grow as much as 17% globally as its growing region is being expanded into higher latitudes with a warmer climate. For Europe, the projections are similar[5]: corn yields are expected to decline by up to 22%, while wheat yields in Northern Europe could experience some productivity gains as the ideal growing climate shifts north. Yields could decrease by 49% in Southern Europe (mostly due to the limited availability of water that the region is expected to experience under a changing climate).

Will Climate Change Benefit Some Crop Yields?

For the most part, the impacts of climate change are expected to affect the yields of crops negatively, but this is not always the case. The process of photosynthesis uses CO_2 and water to make energy, which begs the question: Does this mean that increased levels of CO_2 will translate into more photosynthesis, resulting in higher yields? The answer is yes and no.

As with soils, there are different types of photosynthesis. Most plants use C3 photosynthesis (including rice, wheat, oats, barley, cotton, soybeans), while C4 plants (which include corn, sugarcane, sorghum) are a unique minority better adapted to higher temperatures and drier climates. What is important is that C3 plants are limited by CO_2 availability, which means that higher CO_2 concentrations in the atmosphere will result in more growth and yield,[6] albeit in a limited way. In contrast, higher CO_2 concentrations do not have any significant effects on C4 plant yields because this type of photosynthesis is not limited by CO_2 levels. As temperatures rise in northern latitudes, the area for C3 plants (such as wheat) to thrive in will expand, which comes on top of the higher CO_2 concentrations in the atmosphere being favorable for them. This can partially explain why wheat yields are expected to increase in the coming years, though these benefits are not expected to last.

Under higher CO_2 concentrations, nitrogen now becomes the limiting factor for C3 plant yield growth. To sustain this increased growth, higher soil nitrogen concentrations will likely be needed, most likely in the form of fertilizers, the synthetic form of which is neither completely sustainable nor free of negative side effects, especially if used improperly.

These impacts are highly dependent on geography, but local effects will resonate globally. Because of the industrialized nature of our agriculture system, disturbances that affect certain locations can ripple through the global market. For example, cash crops in the US (corn, soybean, wheat) are currently taking a hit due to extreme drought and heat, pushing up global wheat prices by 12% and corn prices by 11%.[7] In the EU, a study found that in the last 50 years, severe heatwaves and droughts have caused crop losses to triple. This highlights the vulnerability of our food system to climate change, and why adaptation is needed to make it more resilient.

---

[5] JRC (2020). *Analysis of climate change impacts on EU agriculture by 2050.*

[6] Taub, D. (2010). *Effects of rising atmospheric concentrations of carbon dioxide on plants.* Nature Education Knowledge, 1(8).

[7] Farm Policy News (2021). *Drought Continues to Impact U.S. Crops.*

GHG Source and Sink

The question of adaptation is particularly relevant for the farming and forestry sectors because they are both a source of and sink for greenhouse gas (GHG) emissions. In the EU-27, farm-gate emissions are responsible for 11% of total emissions (excluding Land Use, Land-Use Change and Forestry (LULUCF)). The agriculture industry is also a major source of non-CO_2 GHG emissions, such as methane (CH_4) and nitrous oxide (N_2O), which have a higher global warming potential.[8] From a sector standpoint, agriculture has always emitted the largest share of non-CO_2 GHG globally as well as in the EU. In the EU in 2020, for example, agriculture was the top emitter, with 383 m tons CO_2eq, followed by energy (149 m tons), waste (150 m tons), and industry (27 m tons). Approximately 59% of EU methane emissions are anthropogenic (caused by human activity), with agriculture topping the list with a share of 53%, emitted by livestock (enteric fermentation, 80.7%), manure management (17.4%) and rice cultivation (1.2%). Meanwhile, N_2O emissions result mostly from agriculture as well as from the (improper) use of mineral and organic fertilizers.

Perhaps the most unique aspect of the farming and forestry sectors is that they are not only a significant source of GHG emissions, but also a natural sink for them, meaning that both can remove emissions from the atmosphere. Land use, land use change and forestry (LULUCF)—referring to how we use and manage land (whether it is for agriculture, forestry or settlements)—largely determines the amount of emissions released into or removed from the atmosphere.

To best understand the concept of a carbon sink, it helps to look at the (terrestrial) carbon cycle (Fig. 7.1). Starting with photosynthesis, plants take in carbon dioxide (CO_2) and water and use it to produce sugar and oxygen. Once this CO_2 is captured by plants, it is then stored in their biomass (leaves, stem, roots, etc.). Biomass is the first place in which carbon can be stored over the long term, like forests or grasslands. Once carbon is in plant biomass, it can then transfer to the soil via the decomposition of dead matter which then—in healthy soils—builds up soil organic matter (SOM) or, via nutrient exchange in the roots, new plants and trees.

Soil is thus the second place that can store (i.e., sequester) carbon over the long term, but not all soils are suitable. The two main types of soil that play

---

[8] Over a 100-year period, CH_4 has a global warming potential of 27.9, while N_2O has a global warming potential of 273, according to the *IPCC AR6*.

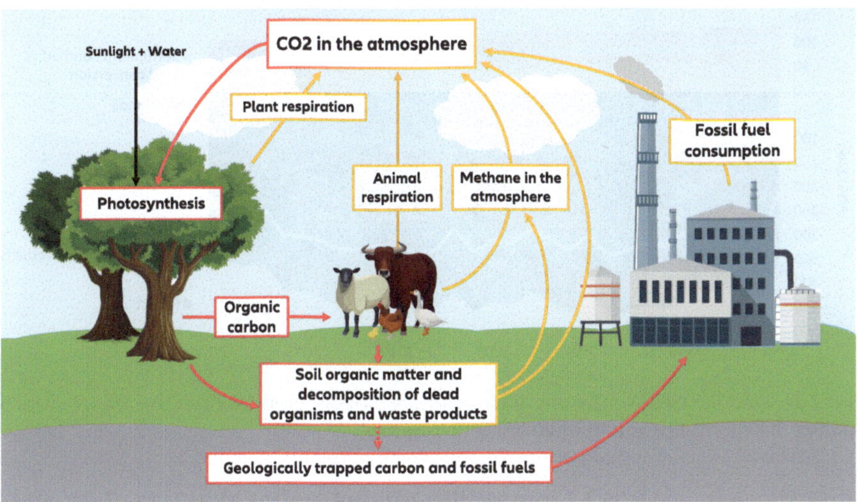

Fig. 7.1   Carbon cycle. Source: Allianz Research

an important role in carbon sequestration are organic and mineral soils. Organic soils, which include peatlands and wetlands, have a high carbon content (at least 20%) and are thought to cover 8% of the EU land area. Their waterlogged conditions prevent microbial decomposition; therefore, the soil's organic carbon is not released into the atmosphere. But they are currently under threat of being drained to use for settlements or turn them into agricultural land. Once drained, microbial decomposition is activated, and the soils begin to release the carbon stored in SOM into the atmosphere. These drained soils are estimated to emit approximately 5% of the EU's total current GHG emissions.

The second soil type, mineral soil, is generally what is thought of as agricultural land. Mineral soils have carbon content below 20% and are known as the "black gold" of farming: high SOM and carbon content means healthy soil, which translates into healthy plants and higher yields. Soils poor in SOM and carbon require the addition of synthetic fertilizers to make up for the lack of nutrients. Unfortunately, most mineral soils have low SOM, resulting from years of unsustainable farming practices, which reinforces the dependence on synthetic fertilizers. In the EU around 7.4 m tons of carbon are estimated to be lost from mineral cropland soils each year.

In 2020, LULUCF was responsible for around 107Mt CO_2e emissions, but it also removed a total of 337Mt CO_2e, resulting in a net sink effect of 230Mt CO_2e (i.e., LULUCF acts as a net sink of 7% of the EU's total GHG emissions). Forest coverage is the EU's largest natural sink, having removed

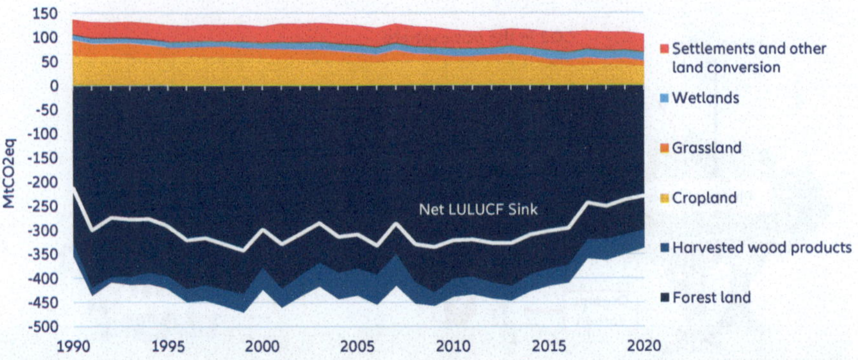

Fig. 7.2 LULUCF GHG emissions and carbon removals 1990–2020 in the EU-27. Source: Allianz Research, European Environment Agency

approximately 301MtCO$_2$ out of the atmosphere in 2020 (Fig. 7.2), and is largely responsible for the net-sink result each year.

Unlike wetlands and conversion for settlements, emissions across grassland and cropland have shown a continuous decrease since the beginning of this millennium. However, this could change as the demand for biomass to produce bioenergy increases, with an expected doubling to 300 Mtoe until 2050. By sector, bioenergy demand will primarily be driven by power generation and residential heating, while demand from the transport sector is not expected to exceed 20% of the total. According to the EU's Fit for 55 plan, 93% of all bioenergy demand will be met domestically, adding to supply security and independence from imports.

On the forest front, while production has increased by about 200 m cubic meters since 1990, no significant impact has been observed on forests' role as a carbon sink. The slight decline in this role witnessed in recent years is thought to be due to pests, wildfires and more intensive harvesting activities. Currently, forest materials (wood and residues) account for only 30% of bioenergy feedstocks, and their share is expected to decrease to 19% by 2050 (although their absolute volume will continue to increase, from 45 Mtoe in 2015 to 65 Mtoe in 2050). The feedstocks that will expand the most include waste, agricultural residues and lignocellulosic grass (Fig. 7.3), while food crops are expected to decline (both in absolute and relative values) to only 2% of feedstock by 2050.

The growth of lignocellulosic grasses as a bioenergy feedstock is welcome. Part of what is known as *energy crops*, lignocellulosic grasses refer to plant dry matter once they have been cut. Among their advantages, they can be grown on a wide range of lands, including those in poor condition and unsuitable for

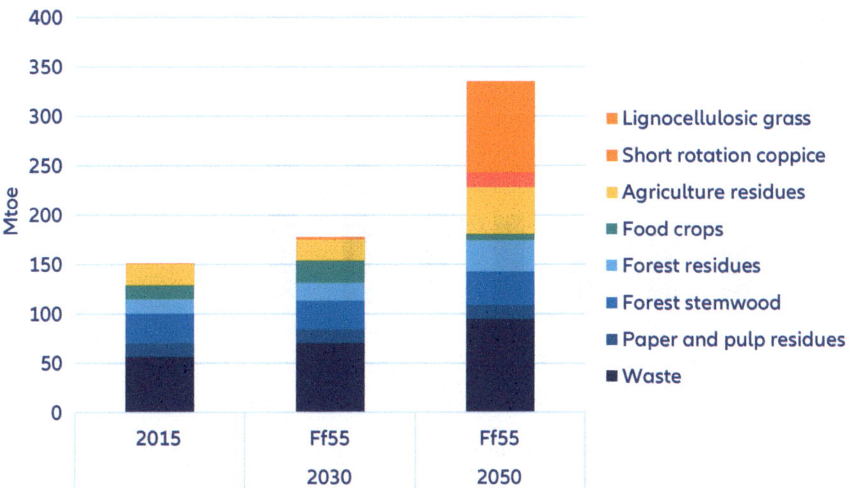

Fig. 7.3 Breakdown of bioenergy feedstocks. Sources: Allianz Research, European Commission

food crops, they are better at preventing soil erosion, and require fewer inputs than row crops. The most beneficial aspect, though, is their potential to become a net carbon sink instead of currently being a net emitter. Trees store most of their carbon in their biomass above ground, while grass stores it underground in its roots. Therefore, grasslands can play a role as a carbon sink and bioenergy feedstock seamlessly and simultaneously, and perhaps even perform better than forests in regions with increased fire risk. A study by the University of California at Davis found that, as carbon sinks, grasslands performed better and were more resilient than forests in California because they are less affected by droughts and wildfires, which are expected to increase in frequency and severity because of climate change.

In the EU, the use of grasslands rather than forests as carbon sinks is especially applicable in the Mediterranean region, where fire hazards have increased over the past few years. As the climate changes, however, all EU countries could benefit from grassland as carbon sinks, since fire risk increases with rising temperatures and more frequent droughts.

Improving the Sink Capacity

If the EU's is to achieve its climate neutrality goals, it will need to do a substantial amount of carbon dioxide removal (CDR), through both nature-based and technological approaches. By 2050, at least 424 m tons of CO_2 will

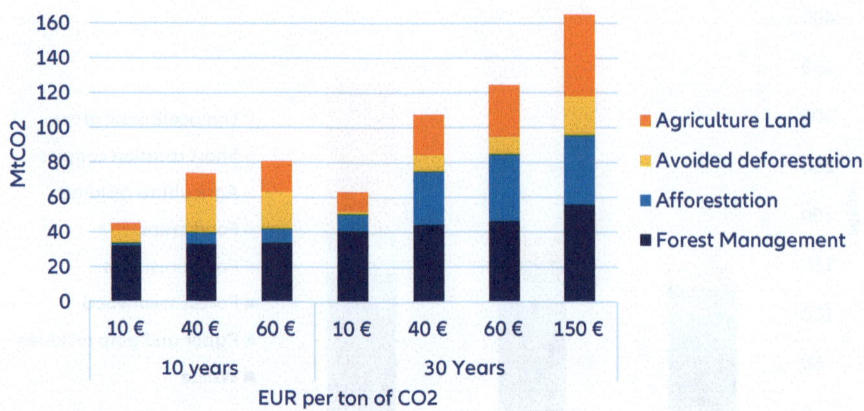

Fig. 7.4 Potential LULUCF sink enhancement, by carbon price. Source: Allianz Research, European Commission

need to be removed *each year* by the land-use sector, which is far above the current sink volumes (a low of 249 tons in 2019).

Given that forests are among the largest natural carbon sinks, forest resource management will play a key role in influencing the overall sink function. In an analysis of LULUCF sink enhancement, a carbon price of EUR60 per ton of CO_2 would result in a sink enhancement of almost 80 m tons by 2030, split between forest management (34 Mt), avoided deforestation (21 Mt), sequestration in agricultural land (17 Mt) and afforestation (8 Mt). By 2050, assuming that carbon price remains at EUR60 per ton, the total sink enhancement would reach 124 Mt., most notably from new, mature forests that were planted by 2030, in addition to the existing old-growth forests (Fig. 7.4). This could enhance the EU MIX scenario[9] forest carbon sink in 2050 from 279 m tons to 403 m tons, but it is still short of the 424 m tons needed.

The Annex to this chapter includes the underlying analysis, which shows the potential for enhancing the LULUCF sink at different carbon prices in the long run. With a CO_2 price of EUR70, the total LULUCF sink could already exceed 130 $MtCO_2$. A CO_2 price of EUR150 in 2050 could increase the forest sink by 118 $MtCO_2$ and the total LULUCF sink by 165 $MtCO_2$, compared to a situation without a CO_2 price for the LULUCF sector. While these sinks are relatively large compared to EU emissions by 2050, they are

---

[9] The EU MIX is included in the EU assessment of the EU Ff55 proposal and is commonly used as the representative scenario for the expected EU Ff55 pathway. See also European Commission. *Climate Target Plan 2030.*

small compared to current emissions, underscoring the need to reduce emissions first.

The greatest potential lies in optimizing forest management practices (changes in stand rotation length, ratio of thinning to final harvest, harvest intensity or harvest locations), which could boost the forest sink by 56 $MtCO_2$. Improving agricultural practices to store more carbon in the soil, in turn, would increase the LULUCF sink by an additional 47 $MtCO_2$. Converting about 5 m hectares of land to new forests by 2050 could remove 40 $MtCO_2$ from the atmosphere annually, for which afforestation and reforestation subsidies would likely be required. Finally, avoided deforestation can add another 22 $MtCO_2$ to the annual carbon sink potential.

For agricultural land, several strategies across the EU are currently working together to reduce emissions and enhance the carbon sink potential, such as the Soil Strategy for 2030, Farm-to-Fork Strategy, Biodiversity Strategy for 2030, and the Common Agricultural Policy (CAP) framework.

The EU Soil Strategy for 2030 was introduced in 2021 to contribute to the objectives of the EU Green Deal, as well as helping the Ff55 proposal of achieving a net GHG removal of $310MtCO_2$eq per year. Some other objectives of this strategy include combating desertification through soil restoration, reducing net conversion of land for settlements to zero by 2050, lowering nutrient losses and chemical pesticide usage by 50% by 2030, and achieving land-use-based net climate neutrality by 2035. Sustainable soil management practices will play a key role in agriculture, such as minimizing soil erosion, building up SOM, mitigating soil compaction and improving soil-water management.

The Farm-to-Fork Strategy, for its part, outlines how the EU will overhaul its food system from farming to consumption. It includes specific 2030 targets, such as reducing chemical pesticides and nutrient losses by 50%, cutting fertilizer use by 20%, decreasing the sale of antimicrobials[10] for farmed animals by 50%, and increasing the share of organic farming to at least 25% of agricultural land. When it comes to consumers, the strategy aims to develop a sustainable-food labelling framework that addresses the nutritional, social and environmental aspects of food products, as well as proposing legally binding targets to reduce per capita food waste by 50% at the retail and consumer levels.

The 2030 objectives of the EU Biodiversity Strategy, in turn, include establishing protected areas for at least 30% of land and sea in the EU; promoting biodiverse landscape features, and halting and reversing the decline of

---

[10] Which include antibiotics, antivirals, antifungals and antiparasitics.

pollinators (such as through reducing pesticide use by 50%); restoring at least 25,000 km of EU rivers to a free-flowing state, and planting 3bn trees.

It is worth bearing in mind that land conversion (for example, for settlements), deforestation and current practices in livestock farming pose larger threats to European biodiversity than climate change does.[11]

Although many of these strategies and targets overlap and interplay, there are several adaptive practices at the farm level that will help the agriculture and forestry sector weather the impacts of climate change (Table 7.1).

Table 7.1 Benefits of alternative, more adaptive and sustainable agriculture/forestry practices

Adaptations	Description	Benefits
Crop rotation	Growing different types of crops each year on the same land.	Enhances biodiversity and soil nutrient management (reducing use of fertilizer and pesticides).
Cover cropping/ mulching	Crops that are planted to cover the soil between growing seasons of a cash crop.	Enhances biodiversity and soil nutrient management (reducing use of fertilizer and pesticides), prevents soil erosion, can attract beneficial insects such as pollinators.
Polyculture/ intercropping	Growing different types of crops on a single piece of land in close proximity, for example planting rows of different crops side-by-side.	Reduces use of pesticides and herbicides by preventing the spread of pests/diseases, increases soil fertility (reducing use of fertilizers), can attract beneficial insects / pollinators.
No or minimum tillage	Minimizing soil disturbance by leaving the crop residue on the soil's surface and, ideally, depositing seeds directly into the soil.	Reduces soil erosion, enhances soil microorganism biodiversity and soil nutrient management (reducing use of fertilizers), increases soil moisture.
Integrated pest management	An ecosystem-based strategy that focuses on the long-term prevention of targeted pests by using biological control, habitat manipulation and modifying practices. Seeks to promote beneficial organisms and target only harmful pests.	Enhances biodiversity (avoids or reduces use of pesticides, insecticides).

---

[11] European Commission. *The Habitats Directive*.

Table 7.1 (continued)

Adaptations	Description	Benefits
Agroforestry	Intentional integration of trees/shrubs into crop and animal farming systems.	Enhances soil nutrient management (more efficient nutrient recycling by trees), reduces soil erosion and nutrient leaching, enhances biodiversity (attracts pollinators), increases diversification.
Organic farming	Using organic (non-synthetic) fertilizers and inputs.	Reduces use of synthetic pesticides, herbicides and fertilizers, builds up soil organic matter and carbon, making soil more resilient and fertile.
Precision farming	Using on-farm modern technology (such as GPS data and tools for precision navigation) to increase the efficient use of inputs like water, fertilizers and pesticides.	Increases efficiency of inputs, reduces use of pesticides, herbicides and fertilizers, reduces waste.
Use of adapted crops	Introducing crops that are expected to be more resilient to changing climate conditions. In forestry, examples include searching genetic pools for species that are resistant to certain pests or more resilient to drought stress.	Reduces the impact of extreme weather risks, can increase biodiversity and genetic diversity of species.
Close-to-nature silviculture or climate-smart silviculture	Promotes natural or site-adapted tree species, mixed forests, diverse stand structures, natural regeneration, clear-cut avoidance.	Increases forests' resilience to climate change impacts (drought, fire, extreme weather events), increases productivity, enhances biodiversity, decreases susceptibility to pests and diseases compared to "plantation" or even-aged silviculture.

The Power of Soil Microorganisms

In modern agriculture, the two most common mineral fertilizers are those providing nitrogen and phosphorous, both of which are considered limiting nutrients, meaning that their availability plays a direct role in plant growth and yield size. But their application is not always beneficial, since mineral fertilizers are a significant source of non-CO_2 emissions (especially mineral nitrogen fertilizers as a source for N_2O) and manufacturing them is very energy-intensive. But the good news is that their application can be reduced by relying on the natural relationships between plants and soil microorganisms.

For nitrogen, cover cropping and crop rotation are not only deemed key agricultural practices to boost carbon storage in the soil, but can also lead to higher yields with fewer inputs. How can this be? The answer lies with legumes, a type of plant family that includes beans, peas, lentils, alfalfa, clover and soybeans. Legumes can form symbiotic relationships with soil bacteria called rhizobacteria in order to "fix" nitrogen, which is the most common limiting nutrient to plant growth and yields. There is plenty of nitrogen around us in the atmosphere as N_2, but this form is not accessible for plants to use. This is where nitrogen-fixing bacteria step in: rhizobacteria can take nitrogen from the air and convert it to a usable form for plants (NH_3, ammonia). Normally the bacteria population is too low to maximize nitrogen fixation, but with legumes, they form "nodules" on the legume roots. The legume feeds the bacteria glucose (carbon), while the bacteria feed the legume nitrogen, which is stored in the root nodules. Once the legume is harvested or dies, the stored nitrogen is released back into the soil, available for use by the next crop, such as corn or wheat.

Cover cropping or crop rotations with legumes have demonstrated their ability to lower emissions of CO_2 and N_2O compared to conventional agriculture, which depends on mineral N fertilization. Cover cropping and crop rotation also build soil organic carbon and reduce overall fossil energy inputs into the soil—all while increasing yields.[12] In fact, it was found that in Europe's temperate environments, crop yields were on average 17–21% higher in grain-legume systems than with wheat mono-cropping.

For phosphorus, power is found in mycorrhizal fungi rather than bacteria. In agricultural soils, phosphorus is easily depleted in the root zone. Even when it is present, it is found in very low concentrations and is generally immobile, meaning that it is not taken up by the plants very easily (or efficiently). This can be improved by mycorrhizal fungi, which form symbiotic relationships with all types of plants and trees. In this relationship, the plant exchanges sugar/glucose with the fungi and the fungi provides improved water and phosphorus nutrient absorption. In agricultural soils, 50 m mycorrhizal fungal networks (called *hyphae*) have been found in 1 g of soil; their large networks offer plants an extended "reach" into areas that would otherwise be inaccessible to the plant. The benefits of mycorrhizal fungi are plentiful, but soil disturbances, such as

---

[12] Stagnari, F., Maggio, A., Galieni, A. et al. (2017), *Multiple benefits of legumes for agriculture sustainability: an overview.* Chem. Biol. Technol. Agric. 4, 2. https://doi.org/10.1186/s40538-016-0085-1

conventional tillage, destroy fungal networks and diversity,[13] which is why it is important to keep such disturbances to a minimum to allow these natural relationships to flourish.

In addition to implementing new agricultural practices, the role of technology is also increasing for farm and forest managers. For example, in forestry, decision-support systems (DSS) help forest managers perform adaptation, mitigation and risk analysis for their forests. DSS help managers make informed decisions about how climate change will impact their operations, including what types of tree species would perform best under the expected future climate conditions for their sites. In both forestry and farm operations, drones or unmanned aerial vehicles (UAVs) are also gaining popularity. UAVs are an innovative risk-management tool for forestry, as they can quickly detect and map damages to forests (either from pests or weather disturbances). Other technologies in forestry also include camera sensors, which are used for the early detection of forest fires.

Similar technologies are also being applied to farming. Precision farming, mentioned above, uses technological solutions to monitor crop growth and field conditions such as moisture and nutrient levels. For example, the use of crop and optical sensors can help apply fertilizers or other inputs more effectively, targeting specific areas where it is needed (reducing widespread application), as well as maximizing uptake. Remote-sensing technology is also a rapidly expanding practice in agriculture, whereby drones are not only used to map land, but also to detect signs of malnourishment or drought even before physical signs begin to show on plants.

Reducing Emissions

To be compatible with staying below 1.5 °C warming, increasing carbon sinks is not enough. Emissions reduction—and in particular of non-CO_2 GHGs—is also required. In fact, starting in 2031, LULUCF carbon sink calculations will also include non-CO_2 emissions from agriculture (CH_4, N_2O). To incentivize and drive down emissions of non-CO_2 gases, a climate neutrality target for farming and forestry by 2035 was introduced for all GHG emissions, which includes LULUCF and non-CO_2 agricultural emissions. Figure 7.5 shows the development paths in the EU baseline scenario (BSL), the EU MIX (MIX) scenario and the EU PLUS scenario (+), representing the pathways for enhancing carbon sinks to a 1.5 °C compliant level.

---

[13] Zahangir Kabir. (2005). *Tillage or no-tillage: Impact on mycorrhizae. Canadian Journal of Plant Science.* 85(1): 23–29. https://doi.org/10.4141/P03-160

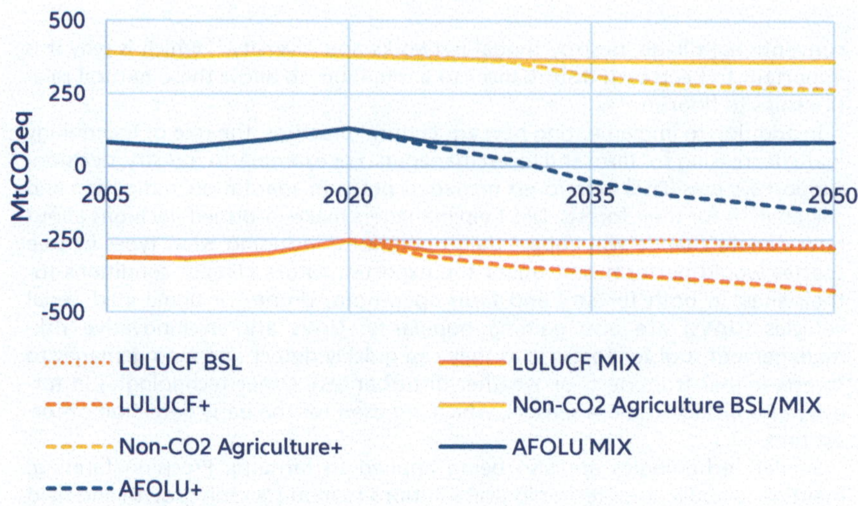

Fig. 7.5 AFOLU+ emissions in the EU MIX scenario*. *The AFOLU line is the sum of the non-CO$_2$ agriculture emissions as in the MIX scenario plus the LULUCF sink projected without additional incentives to enhance the LULUCF sink in MIX. AFOLU+ includes additional action to enhance the LULUCF sink (LULUCF+). Source: Allianz Research, EU 2030 Climate Target Plan

The BSL scenario sets the starting point to determine the investment requirements, and the EU MIX scenario represents the policy ambition best aligned with the EU Green Deal announcements. The enhanced pathways "Non-CO$_2$ Agricultural+" for emissions, "LULUCF+" for CO$_2$ sinks, and the sum of both represented by "AFOLU+" for net emissions show that removals and emissions would need to cancel out (or reach "no debits" in the EU terminology) just before 2035, and that sink capacity would need to steadily increase to deliver "net credits" in subsequent years.

Central to the current discussion on CO$_2$ removal at the EU level is the design of accounting and trading for associated CO$_2$-removal activities, which would likely require national approaches to setting targets and detailed analysis that take account of the differences in the geographic distribution of removals and emissions, including those from non-CO$_2$ emissions.

The enhanced pathways require additional investments. Figure 7.6 shows the gradual increase in associated investment needs, derived from the EU assessment of marginal abatement costs for agricultural emissions and carbon sink enhancement costs, as described in the Annex. The investments increase from an average of EUR460m per year in 2025–35 to an average of EUR14.4bn per year in 2045–50, with about two-thirds of the investment needs originating from non-CO$_2$ GHG abatement and one third from enhancing carbon sinks.

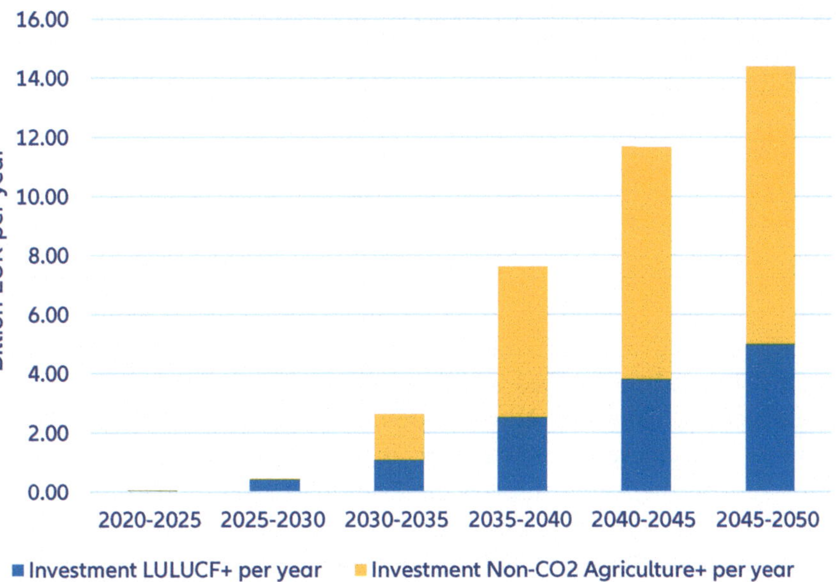

Fig. 7.6 Investment needs for advancing LULUCF carbon sinks and agriculture non-CO_2 abatement beyond baseline ambitions to the AFOLU+ goals. Source: Allianz Research

The financial incentive policy instruments designed to change practices in forestry and agriculture would mean that uses for the associated biomass products (wood, pulp and paper, fabrics, advanced biofuels, etc.) would face new economic competition. In addition, the conservation of carbon stocks in the sector (increasing the sink through the avoidance of emissions rather than improving forest management) could be upgraded, with potential positive side effects for biodiversity and the other ecological functions of existing forests. Policies should also consider the resulting risk of changes in supply chains that may stimulate imports and reduce the global environmental benefit as well as the EU's economic and social benefits.

Table 7.2 summarizes the investment needs until 2030, and in the long run, from 2030–2050. Notably, the long-run perspective reveals an investment gap of around EUR8.3bn per year. The difference between the EU's Ff55 ambition (referring to the MIX scenario) and the 1.5 °C target (referring to the "+" scenario) is striking.

Particularly at high carbon prices, agriculture is the sector with the second-highest abatement potential. Figure 7.9 in the Annex illustrates this potential.

Clear win-win investment opportunities include biogas recovery for dairy cows and cattle farms. Its use would also allow an increase in the supply of biomass available for biomethane production. Breeding selection can also

Table 7.2 Investment needs for advancing LULUCF carbon sinks and agriculture non-CO_2 abatement beyond baseline ambitions to the AFOLU+ goals

Investment in billion EUR	2020–2030	2030–2050
Total investment MIX	2.41	7.61
Total investment MIX per year	0.24	0.76
Total investment+	2.67	181.84
Total investment+ per year	0.27	9.09
Investment gap MIX vs. +	0.26	174.23
Investment gap MIX vs. + per year	0.03	8.33

enhance productivity, fertility and longevity to minimize the methane intensity of dairy and meat. Feed additives and feed-management practices can reduce methane emissions. Nitrification inhibitors and technologies allowing for more efficient fertilizer use offer an option to reduce nitrous oxides. Figure 7.10 in the Annex displays the main investment opportunities in the LULUCF sector. They include—ordered by decreasing contribution—the enhancement of the forest sink, optimizing forest management practices, improving agricultural practices to store more carbon in the soil, incentives for additional reforestation, and avoided deforestation.

The Role of Carbon Markets

Carbon markets, where emission allowances are traded, are emerging as a key player on the path to climate neutrality. There are two types of markets: regulatory compliance and voluntary markets. Compliance markets are used where by law, companies or governments must account for their emissions. The voluntary carbon market reached a new record value of about USD2bn in 2021 globally, and could potentially reach USD40bn in 2030.[14] That is still small compared to the USD850bn of carbon allowances traded in compliance markets in 2021.

A prominent example of a market that is used for compliance purposes is the Clean Development Mechanism (CDM), which was established by the Kyoto Protocol. This mechanism allows countries to fund greenhouse gas emissions-reducing projects in other countries and claim the saved emissions as part of their own efforts to meet international emissions targets. For industrialized countries, an emission-reduction project must take place in a developing country; after completion, a carbon credit called "Certified Emission Reductions" (CER) is generated. Most projects are focused on renewable

---

[14] See also: Shell and BCG (2022). *The voluntary carbon market – 2022 insights and trends.*

Table 7.3 COP26 key takeaways for Article 6

Article	Confirmed Provisions
6.2	Provides for "cooperative approaches" to be established directly between countries for the purpose of trading Internationally Traded Mitigation Outcomes (ITMOs) units, with corresponding adjustments required for any ITMO transfers made.
	Like carbon credits, these ITMOs are where emission reductions are passed from one country's GHG account to that of another country. This can happen at both the government and corporate level. Double counting is avoided by applying a corresponding adjustment, where the "host country" (the country where the carbon reduction is located) must first approve the transfer, then adjust its own GHG account, before it being credited to another country.
6.4	Provides a "top-down" global platform for trading of ITMOs by all countries, operated by UNFCCC. This "Sustainable Development Mechanism" will replace the current "Clean Development Mechanism (CDM)" introduced under the Kyoto Protocol. A Supervisory Body and secretariat housed within UNFCCC will provide oversight.
	Activities that generate emission reduction units are defined as emission reductions or removal activities and may not represent activities conducted before 2021.
6.8	Provides for the facilitation and collective coordination of non-market approaches to be taken by countries to drive emissions down. These can engage both country and non-country actors. Examples include pledge programs, initiatives, statements, roadmaps and commitments that collaboratively support emission reduction activities.

Source: Allianz Research

energy or energy efficiency, while projects in the agriculture, forestry, and other land-use sectors are somewhat more restricted. The CDM has faced several challenges and showed only limited success in reducing GHG emissions.

A major success of the COP26 in Glasgow was to finalize and agree on the formulation of Article 6 of the Paris Agreement (see Table 7.3). It builds on the aims of the CDM, addresses its deficits, and provides a framework for implementing a global carbon market that includes the scaling-up of natural carbon sinks and trading of the resulting carbon offsets. The institutions related to and the markets based on Article 6 have to be established within the coming years to allow the international linkage between offset activities and offset financing. In the process, the CDM will be replaced by the new SDM (Sustainable Development Mechanism).

Voluntary carbon markets have emerged as a complement to compliance markets, especially where companies have made collective agreements within their industries to address their emissions. These markets trade carbon credits on a voluntary basis. The carbon credits awarded are called "Verified Emission Reductions" (VER), which are then registered at a specific carbon registry, which certifies the results. There are four main registries, or standards, for these carbon offsets: Verified Carbon Standard or Verra, Gold Standard, American Carbon Registry, and the Climate Action Reserve.

In both markets, certification standards ensure that the four core principles of carbon finance are adhered to: additionality, no overestimation, permanence, and exclusive claim. They also provide additional social and environmental benefits.[15] For the land-use sector, such as projects in forestry or agriculture, permanence is a real concern, since there is a risk that the carbon stored or avoided is lost if a disturbance such as a fire or natural disaster occurs.

One example of a recently established compliance mechanism is CORSIA, the Carbon Offsetting and Reduction Scheme for International Aviation, which will require participating airlines to buy carbon credits for their emissions growth above their 2020 levels. Although this is a compliance scheme, it does allow credits from certain standards on the voluntary carbon market (such as the Gold Standard) as long as the credits meet certain requirements. What is important for the land-sector project, though, is that only certain types of land-use activities are allowed. For example, CORSIA will not accept CDM credits from afforestation or reforestation projects because of the lack of policies to ensure permanence. Furthermore, REDD+ credits (see box) are also not accepted.

Forests Seeing REDD+

In addition to carbon removal, there is also carbon avoidance. Perhaps the most visible are REDD+ projects, which stands for Reducing Emissions from Deforestation and Forest Degradation. These projects avoid carbon emissions by preserving forests in specific areas that are at risk of deforestation.

However, it is particularly difficult to calculate these projects' emissions avoidance potential. First, you must first have a "baseline," which is the amount of carbon that would likely be released if deforestation occurred. With REDD+ projects, the baseline is calculated using the deforestation rate over the past 10 years in a nearby, comparable area. Because there is no absolute certainty, some argue that these projects violate one of the core principles of carbon finance: "no overestimation". But these projects generally have lower costs than afforestation projects and show immediate results (think of preserving a forest that already exists versus rebuilding a forest that takes time to mature).

It is important to note, though, that CDM and Gold Standard exclude avoided deforestation projects and that, as things stand, these credits or activities will not be accepted in the proposed Sustainable Development Mechanism (scheduled to replace the CDM), which sets higher requirements for reduction or removal activities.

---

[15] See Annex I to this chapter for definitions.

In the context of carbon markets, one of the most intriguing ideas is the EU's plan to boost the carbon sink via carbon farming. This new initiative aims to develop and deploy nature-based carbon-removal solutions at scale across the EU using farming and forestry projects. To drive this business model and develop incentives to boost carbon storage and removal via the land sector, the EU is discussing a regulatory framework for the certification of carbon removal, in which the role of tradeable carbon credits wasn't clear by the middle of 2023. This could essentially create an EU-regulated carbon market where farmers—or land managers, in the new terminology—can be awarded carbon credits by the EU, which they could then sell in the carbon market at their own discretion. These credits could become a new source of income for farmers and further incentivize them to adapt more climate-friendly practices. Meanwhile, the consultations around EU carbon farming are still ongoing and the implementation is still largely unspecified.

After recognizing that private schemes and voluntary markets vary in transparency, integrity and quality, the EU decision to establish its own scheme to standardize the methods and rules for monitoring, reporting and verifying (MRV) carbon credits with scientifically robust requirements could provide an important benchmark for the global market. The objective is for each land manager to have access to verified emission and removal data by 2028, with the potential for carbon-farming initiatives to remove at least 42Mt CO_2eq per year by 2030 (contributing to the EU's overall net carbon dioxide removal target of 310Mt CO_2eq in the LULUCF sector).

The EU highlights the following practices as having the most potential in carbon farming:

1. Afforestation and reforestation
2. Use of conservation tillage and cover crops in agriculture to build soil organic carbon in mineral soils
3. Restoring, re-wetting and conserving peatlands and wetlands
4. Targeted conversion of cropland to fallow or permanent grassland
5. Agroforestry and other types of mixed farming

The EU's REFORMED COMMON Agriculture Policy: Is It Enough?

The latest EU reform to its Common Agricultural Policy (CAP) allotted an annual budget of EUR386.6bn for the current funding period (2021 to 2027). At 26%, the CAP is the single largest item in the total EU budget. With its benefits covering around 85% of EU farmland, its reach is far, wide and influential.

The reform has been welcomed as climate-friendlier than the previous CAP, which was criticized for not doing enough to prevent climate change: An audit from JRC found that it did not contribute significantly to emissions reductions. Considering this, the new reform has boosted funding and introduced several green-tinged amendments:

- Introducing eco-schemes. The first pillar of the CAP, which accounts for the bulk of funding at EUR291.1bn, includes the direct payment of funds to farmers based on the size of their operations. From 2023 to 2024, 22% of funding will be dedicated to farmers who practice sustainable activities that contribute to the targets of the EU Green Deal; this will be increased to 25% from 2025 to 2027. This first pillar also comes with enhanced conditionality: to receive the funds, farmers must follow a more stringent set of sustainable farming requirements, the "Good Agricultural and Ecological Conditions." For instance, crop rotation will be required on at least 10 hectares, and, on every farm, at least 3% of arable land is dedicated to biodiversity and non-productive elements, with a possibility to receive support via eco-schemes to achieve 7%. Wetland and peatlands must also be protected.
- Increased funding for rural development, traditionally the second pillar of funding (EUR95.5bn). At least 35% of this budget is assigned to environmental and climate measures. This pillar also includes research and development support in the form of advisory services, knowledge exchange and training actions, an important resource to boost carbon farming.
- Operational programs. For fruit and vegetable production, these programs must dedicate at least 15% of their funds to the environment (+5 pp. compared to current program).
- CAP strategic plans: To receive EU funding during the next period, member states had until the end of 2021 to submit their own CAP strategic plans detailing how they plan to implement these rules.

These reforms are aimed at boosting funding to drive the previously mentioned strategies (biodiversity, Farm-to-Fork, soil strategy and many others) and overall achieve the proposed Ff55 targets, with an expected 40% of the overall CAP budget contributing to environmentally friendly actions.

Whereas previously the CAP focused on action-based results, rewarding farmers for implementing certain practices in their operations regardless of the result achieved, the new direct payments incentivize farmers to invest in their operations and to transition to sustainable agricultural practices, especially via the eco-schemes. Moreover, to incentivize this further, a results-based approach will be introduced via the carbon-farming initiative and carbon-removal certification system, which was still in discussion at the time of printing in mid-2023.

The Demand Side

Agricultural and land-use policy is not concerned solely with the emissions associated with the farm-to-gate production side. In its farm-to-fork perspective, it also includes the emissions further down the value chain, including the effects of consumption behavior changes, like switching to a vegetarian diet. Globally, the FAO estimates that 31% of anthropogenic emissions came from the agri-food systems in 2019. In the EU-27, agri-food system emissions accounted for 32% of total GHG emissions. Of this share, most of the emissions (51.1%) originated from pre- and post-production activities, followed by farm-gate emissions, at 45.5%, and lastly, land-use change, at 3.4% (Table 7.4).[16]

A behavioral change towards a more sustainable, low-emission dietary choice is important as well. Food products differ significantly with regards to

Table 7.4 EU-27 emission sources across the agricultural value chain, defined by FAO

LULUCF (without carbon removal) (3.4%)	Forest conversion to other land uses
	Peat fires
	Drained organic soils
Farm-Gate (45.5%)	Crop residues (including burning)
	Drained organic soils
	Enteric fermentation (E.g., livestock)
	Manure management (including soil application and pasture)
	Rice cultivation
	Synthetic fertilizers
	On-farm energy use
Pre- and Post-Production (51.1%)	Food transport
	Processing & packaging
	Refrigeration
	Retail & cooking
	Fertilizer manufacturing and other pre-production
	Solid food waste
	Incineration
	Wastewater (industrial and domestic)

Source: Allianz Research, FAO

---

[16] The FAO definitions deviate from the EU ones used above, but they are still in the same ballpark. According to FAO, roughly 16% of total EU GHG emissions result from farm-to-gate and LULUCF (49.9% of 32% of total GHG emissions from the agri-food system). In our EU analysis 11% result from non-CO_2 agriculture plus about 4% from LULUCF if negative emissions are disregarded.

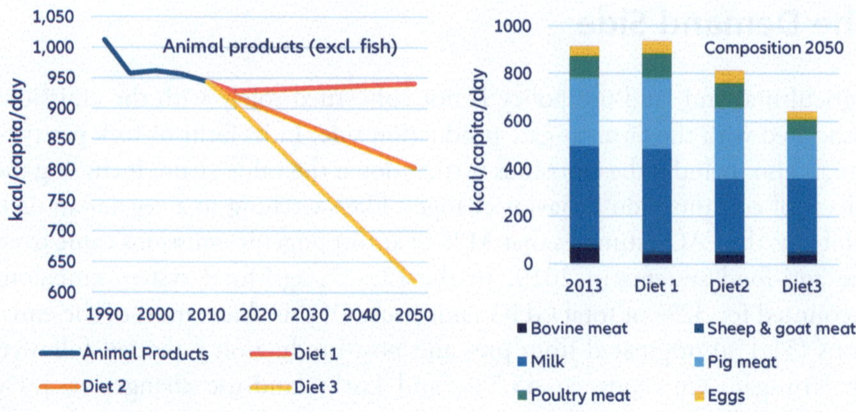

Fig. 7.7 Evolution of consumption and composition of animal products for different dietary choices. Sources: Allianz Research, European Commission

GHG emissions and energy consumption during their production and transportation. Red meat is an example of a GHG-intensive dietary choice that comes to mind immediately. It is often resource- and energy-intensive and contributes directly to methane emissions. But also fruits and vegetables that must be transported over long distances, or cooled for non-seasonal consumption, can be rather GHG-intensive.

Figure 7.7 depicts the evolution of total consumption and the composition of animal products for three different dietary choices, ranging from light decreases in meat and dairy (Diet 1) to more substantial decreases (Diet 3). Diet 3 is consistent with levels of meat consumption recommended in a number of diet studies.[17] All dietary scenarios are also in line with the UN Sustainable Development Goal of halving per-capita food waste generation at the retail and consumer levels by 2030.[18] These diets would additionally bring with them health benefits, though in all diets dairy and meat consumption would still remain at a relatively high level.

The moderate changes in food consumption patterns could significantly reduce emissions from farming production. The effect in 2050 ranges from 34

---

[17] AgCLIM50 Project (2017). *Challenges of Global Agriculture in a Climate Change Context by 2050*, Bajželj, B., Richards, K., Allwood, J. et al. (2014). *Importance of food-demand management for climate mitigation*. Nature Clim Change 4, 924–929. https://doi.org/10.1038/nclimate2353 and Bryngelsson et al. (2016). *How can the EU climate targets be met? A combined analysis of technological and demand-side changes in food and agriculture*. Food Policy, Volume 59, Pages 152–164. https://doi.org/10.1016/j.foodpol.2015.12.012

[18] United Nations (2015). *Transforming our world: the 2030 Agenda for Sustainable Development.*

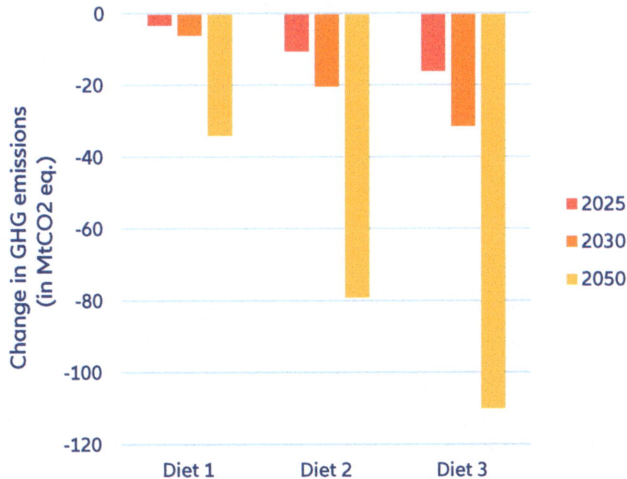

Fig. 7.8 Greenhouse gas emissions effects of different dietary choices through 2030. Source: Allianz Research, European Commission

$MtCO_2eq$ with Diet 1, to 110 $MtCO_2eq$ with Diet 3 (Fig. 7.8). The transition would continue after 2050, aiming for full implementation by 2070. In terms of emissions reduction, the behavioral change would be within the same order of magnitude as the agriculture sector's technical reduction potential.

Conclusion

In the complex landscape of climate change, the symbiotic relationship between forestry, agriculture, food production, and land use emerges as a critical focal point. Both agriculture and forestry are vital economic sectors, encompassing around 39% of the EU's total land area and providing significant global contributions to food production. Simultaneously, they stand as the sectors most susceptible to climatic volatility, with potential adverse impacts on productivity due to changing weather patterns.

The nuanced impact of CO_2 concentrations on various types of plants has illustrated the potential for both gains and losses in crop yields, with geography playing a crucial role. However, the prevailing trend indicates an overall vulnerability of the food system, emphasizing the pressing need for resilient adaptation strategies.

The farming and forestry sectors uniquely act as both sources of and sinks for greenhouse gas emissions. This dual role opens opportunities for strategic management through carbon sequestration in both forests and emerging potentials in grasslands. Still, these areas also present challenges, such as the draining of organic soils contributing to emissions and the balance of biomass demand for bioenergy.

The move towards alternative feedstocks like lignocellulosic grasses, offering potential as carbon sinks, and the continuous decrease in emissions across grassland and cropland is promising. Yet, there is an underlying urgency for thoughtful management of land resources, considering the complexities of soil types, carbon cycles, and global interdependencies.

As climate change continues to reshape the environment, the need for comprehensive, multifaceted strategies that integrate sustainable land management, forestry practices, agricultural productivity, and climate resilience becomes paramount. The European region's response will not only shape its own future but also influence global food security and environmental stability in the decades to come.

The EU's latest reform of its Common Agricultural Policy (CAP) has allocated EUR 386.6 billion for 2021 to 2027, constituting 26% of the total EU budget and benefiting 85% of EU farmland. This reform introduces environmentally friendly amendments to support the EU Green Deal's targets, including eco-schemes, more funding for rural development, and specific programs for fruit and vegetable production. A focus on sustainability will allocate 40% of the CAP budget to green actions, and a shift from action-based to results-based approaches is being emphasized. On the demand side, the significance of agri-food system emissions is recognized. The text highlights the role of sustainable dietary choices in reducing emissions, outlining different scenarios aligned with the UN goal of halving per-capita food waste by 2030. The combined agricultural reforms and emphasis on consumption habits form a holistic approach to reducing the EU's greenhouse gas emissions.

Appendix 1

Core Principles of Carbon Finance

Additionality. The project should not be legally required, common practice or financially attractive in the absence of credit revenues.

No overestimation. CO_2 emissions reduction should match the number of offset credits issued for the project and should take into account any unintended GHG emissions caused by the project.

Permanence. The impact of the GHG emission reduction should not be at risk of reversal and should result in a permanent drop in emissions.

Exclusive claim. Each metric ton of CO_2 can only be claimed once and must include proof of the credit retirement upon project maturation. A credit becomes an offset at retirement.

Provide additional social and environmental benefits. Projects must comply with all legal requirements of their jurisdiction and should provide additional co-benefits in line with the UN's Sustainable Development Goals.

Appendix 2

Calculation of Investment Needs

Table 7.5 shows the investment needs, as determined by the additional emissions that should be avoided or the carbon sink increase relative to the baseline scenario (BSL).

"MIX" is the main scenario we use from the EU assessment for the Ff55 proposal. While MIX is more ambitious than the BSL scenario for the LULUCF carbon sinks, it is similar for the non-CO_2 agricultural emissions.

AFOLU is the sum of LULUCF and "Non-CO_2 Agriculture."

Table 7.5 MtCO$_2$-eq emissions in the EU Baseline, EU MIX and increased-ambition EU PLUS scenario

	2005	2010	2015	2020	2025	2030	2035	2040	2045	2050
LULUCF BSL	-311	-315	-298	-252	-261	-258	-262	-251	-266	-271
LULUCF MIX	-311	-315	-298	-252	-289	-295	-291	-286	-280	-279
LULUCF+	-311	-315	-298	-252	-309	-340	-361	-381	-400	-424
Non-CO2 Agriculture BSL/MIX	393	381	390	383	373	368	365	363	362	363
Non-CO2 Agriculture+	393	381	390	383	379	363	312	283	274	267
AFOLU MIX	82	66	92	131	83	74	73	76	81	83
AFOLU+	82	66	92	131	69	24	-49	-99	-127	-157

Source: EU 2030 Climate Target Plan. "Commission Staff Working Document Impact Assessment" accompanying the document "Communication from the Commission to the European Parliament, the Council, the European Economic and Social Committee and the Committee on the Regions" Stepping up Europe's 2030 climate ambition Investing in a climate-neutral future for the benefit of our people (SWD/2020/176 final)

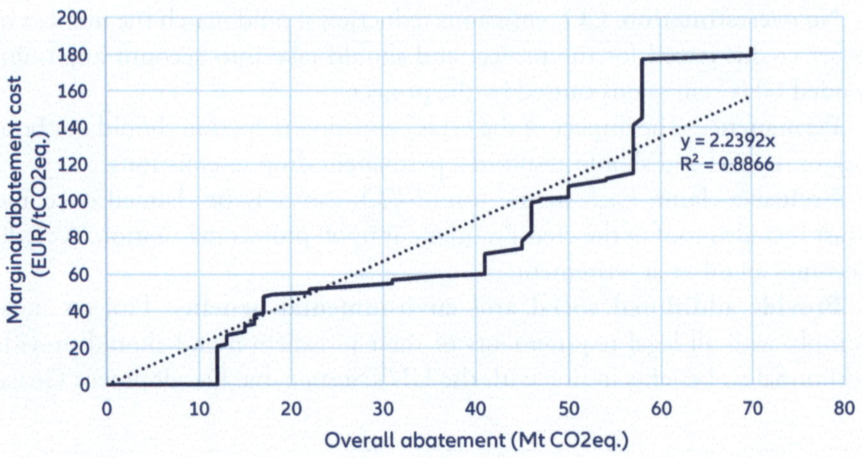

Fig. 7.9 Marginal abatement cost curve for all non-CO_2 agricultural sector greenhouse gas emissions [For the investment analysis, the marginal abatement curve is approximated by $y' = 2.24x$. As this is dominantly composed of infrastructure and process change investments, we approximate the investment cost by the integral of the marginal abatement curve, which is $y = 1.12x^2$ (y: difference baseline vs. enhanced ambition)]. Source: EU 2030 Climate Target Plan. "Commission Staff Working Document Impact Assessment" accompanying the document "Communication from the Commission to the European Parliament, the Council, the European Economic and Social Committee and the Committee on the Regions," "Stepping up Europe's 2030 climate ambition," "Investing in a climate-neutral future for the benefit of our people (SWD/2020/176 final)"

"LULUCF+", "Non-CO_2 Agriculture+" and "AFOLU+" describe the enhanced-ambition pathways from the EU assessment.

Figure 7.9 shows the marginal abatement curve for the non-CO_2 GHG-emissions in agriculture. Approximated linearly for the analysis. Figure 7.10 shows the EU assessment for the marginal potential enhancement of the carbon sink for increasing price levels as the blue line. For the analysis this has been approximated by the second order polynomial shown. As the underlying measures predominantly refer to capital investments and not to switching to processes that imply more operational costs, the total investment needs can be approximated in both cases by the integral under the marginal cost curve, according to economic theory.

Agriculture is the sector with the second-highest abatement potential, particularly at the higher carbon price. Figure 7.9 below illustrates this potential, showing that mitigation options exist at significant price differences. The dotted lines indicate marginal mitigation costs of EUR10/tCO_2eq and EUR55/

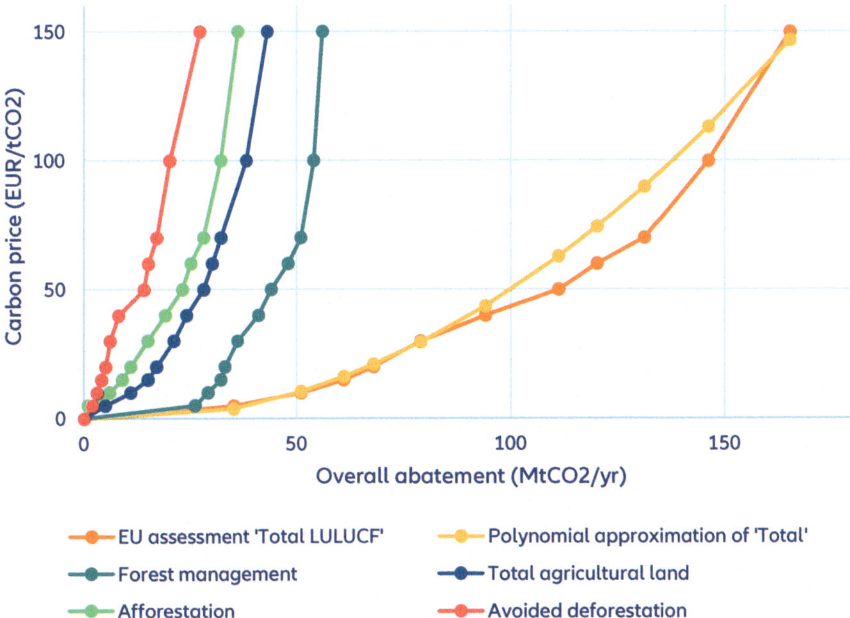

Fig. 7.10 Potential for carbon sequestration and LULUCF sink enhancement at different carbon prices [For the investment analysis, the marginal curve is approximated by the second order polynomial $y' = 0.006x^2 - 0.1x$. Investments are approximated by the (positive part of the) integral of the marginal curve, which is $\max\{0; y = 0.002x^3 - 0.05x^2\}$ (y: difference between baseline vs. enhanced ambition). The investments can be decomposed on request into: Agricultural land, forest management, avoided deforestation and afforestation]. Source: EU 2030 Climate Target Plan. "Commission Staff Working Document Impact Assessment" accompanying the document "Communication from the Commission to the European Parliament, the Council, the European Economic and Social Committee and the Committee on the Regions," "Stepping up Europe's 2030 climate ambition," "Investing in a climate-neutral future for the benefit of our people (SWD/2020/176 final) and "In Depth Analysis in Support of the Commission Communication COM(2018) 773", "A Clean Planet for all A European long-term strategic vision for a prosperous, modern, competitive and climate-neutral economy"

tCO_2eq, respectively reducing emissions by 3% and 8% compared to baseline by 2030. One of the most economical options for clear win-win strategies is farm-scale anaerobic digestion with biogas recovery; it can be an important emission-reduction technology for dairy cows and cattle farms, for both small and large farms. Its use would also help to increase the supply of biomass available for biomethane production, a technology that will see increasing relevance in the future. Selective breeding, both of dairy cows and sheep, could enhance productivity, fertility and longevity, and minimize the methane intensity of dairy and meat products. Moreover, feed additives combined with changed feed-management practices can reduce methane emissions, irrespective of farm size.

Nitrification inhibitors are an option at higher marginal costs for larger farms (30 to 150 hectares) to reduce nitrous oxides at scale. The same applies for variable-rate technology to reduce emissions of nitrous oxide emissions related to more efficient fertilizer use.[19]

Overall, the results show that a significant number of win-win abatement technologies exist for agriculture.

---

[19] Variable rate technologies allow the right quantities of fertilizer to be applied at the right place. This helps to both maintain yields and avoid nitrogen losses. See also Karin Späti, Robert Huber, Robert Finger (2021).

Benefits of Increasing Information Accuracy in Variable Rate Technologies. Ecological Economics, Volume 185. https://doi.org/10.1016/j.ecolecon.2021.107047

8

Africa Unbound

Global warming will be cruelest to Africa, despite being the continent that contributed the least to global warming—bar Antarctica. Even today, Africa emits far less carbon than other continents, and yet, its vulnerability to climate change means that it will be burdened with the highest impacts and losses from it.

An African energy transition would thus be highly beneficial, and would offer the entire continent a chance to leapfrog in its development. It would not only help to mitigate climate change, but also reduce poverty by increasing labor productivity, agricultural yields, water availability, food security and human health. Moreover, while mainly low-skilled jobs in "brown" sectors would be lost, higher-skilled jobs in "green" sectors would be gained, potentially leading to much higher human capital levels. In a nutshell, an African energy transition opens significant development prospects for the whole continent.

But while the case for such an energy transformation is huge, so are the obstacles. In general, investment needs relative to local GDP are larger than in advanced economies. This, together with insufficient local funding resources, will make it very difficult for most countries to finance their energy transition. They will therefore need to attract international capital, but their access to foreign financing is hampered by such factors as ubiquitous political instability and weak rule of law.

Regardless, the time for action is now. Investments in clean energy have declined in recent years in many emerging markets, not least due to the covid-19 pandemic. If this path continues during the coming decades, African emissions

L. Subran, M. Zimmer, *Investing in a Changing Climate*, Professional Practice in Governance and Public Organizations, https://doi.org/10.1007/978-3-031-47172-8_8

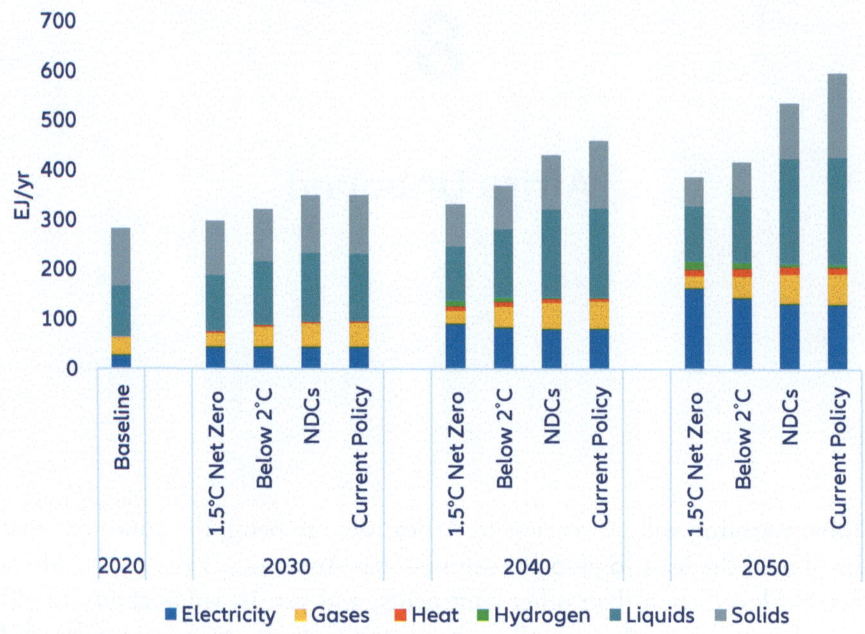

Fig. 8.1 Final energy demand by sector, Africa's 10 largest economies. Sources: NGFS, Allianz Research. Climate scenarios described in Box 1 in the introduction to this book

will grow unchecked and the continent might follow the high-carbon pathways that today's developed economies took in the past, particularly as its final energy demand is set to grow in all scenarios (Fig. 8.1). This would not only undermine global efforts to reduce emissions but would also be a great missed opportunity: the costs of clean-energy technologies have come down steeply in recent years, offering new opportunities for leapfrogging and transitioning to clean energy sources at grand scale without jeopardizing economic aspirations. The future of the African energy sector must be shaped today.

What needs to be done to seize this once-in-a-lifetime opportunity for African development? It boils down to one crucial question: How can the private capital of advanced economies—which, at above EUR200trn, is more than plentiful—be mobilized for investing in Africa's green transition?

The place to start is, clearly, the countries themselves. To attract foreign capital, they need to provide a suitable investment climate by strengthening institutions, improving political governance, and modernizing policies and regulations.

But even these improvements might not be sufficient. From the perspective of private investors, risk-adjusted returns could still be too low. Fortunately, a solution exists: multilateral lenders and development financial institutions could lower risks by taking mezzanine positions in blended finance vehicles, which would make a large number of transition projects more viable for private investors.

Thus, what is really needed is that developing economies in Africa, advanced economies, and private investors join forces in public-private partnerships which represent the most suitable mechanism to foster the African energy transition.

But even if public-private partnerships can be put together, one further challenge remains: reaching the necessary implementation speed. Lofty goals with no clear roadmap for their achievement often amount to naught. Thus, clear green-energy strategies and ambitious yet realistic goals, coupled with economy-wide transition plans and sector-specific pathways, all underpinned by judicious implementation policies, can provide the necessary guidance to unlock private capital.

This chapter will try to shed some light on this issue by providing initial transition plans for the energy sector in the 10 biggest African economies

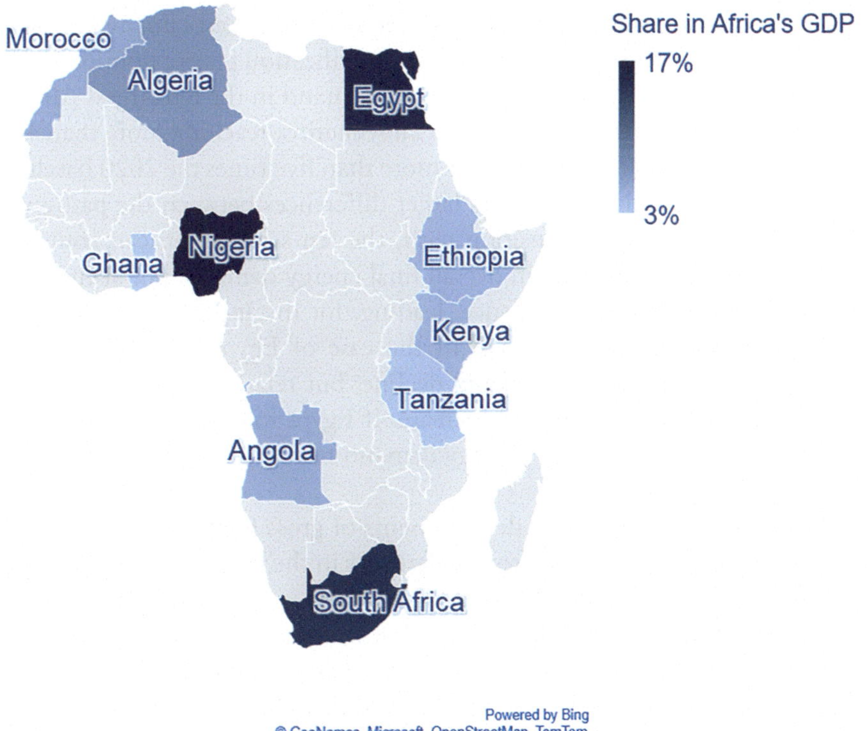

Fig. 8.2 The 10 largest African economies by share in the continent's GDP (%). Source: Allianz Research. 2020 GDP in current USD

(Fig. 8.2), which together represent well over 60% of Africa's total GDP. The purpose is to start an informed discussion based on the required scope and scale of the challenge, with governments and investors welcome to adjust and improve these transition plans—and then walk the talk with actual projects and investments.

No Prosperity Without Growing Energy Demand

The transformation of the energy sector faces a particular challenge in developing economies, such as those in the African continent. As they catch up with more industrialized nations and their population continues to grow, their demand for energy is set to increase significantly, with their investment needs rising apace. At the same time, African economies are already facing considerable impacts from climate change. Hence, any possible investment paths require a thorough assessment for a *sustainable* energy transition. Achieving these targets hinges both on swift action and on the extent to which financing is available from both public and private sectors.

On the demand side, expectations differ across the four climate scenarios (Fig. 8.3). As in high-income economies, electrification plays a dominant role in the African energy transition. Electricity demand in the ten largest African economies (African 10) will increase in all scenarios, reaching more than 100 EJ/year (about 28,000 TWh/year), or more than five times the 2020 baseline. However, aggregate demand shows larger differences between the pathways. Along the 1.5 °C energy-transition path, the ten selected African countries would experience an increase in overall final energy demand, but at the same time reduce their fossil dependence, leading, for instance, to a lower use of solid energy sources such as coal. In the case of Egypt (Fig. 8.4), overall demand for energy from coal would decline, but relative demand for liquid energy sources such as oil would increase. If Egypt follows a path consistent with the 1.5 °C climate target, energy demand from hydrogen would need to increase as well.

The disparities between the developments of gross final energy demand in the scenarios originate mainly from differences in the effort to improve energy efficiency. The 1.5 °C and 2 °C scenarios assume a significant push towards energy conservation, with the introduction of more efficient industrial processes, greener means of transportation and energy savings in both the commercial and residential sectors. The separation of final energy demand by sector is depicted in Fig. 8.1 above for the aggregate of the ten African economies.

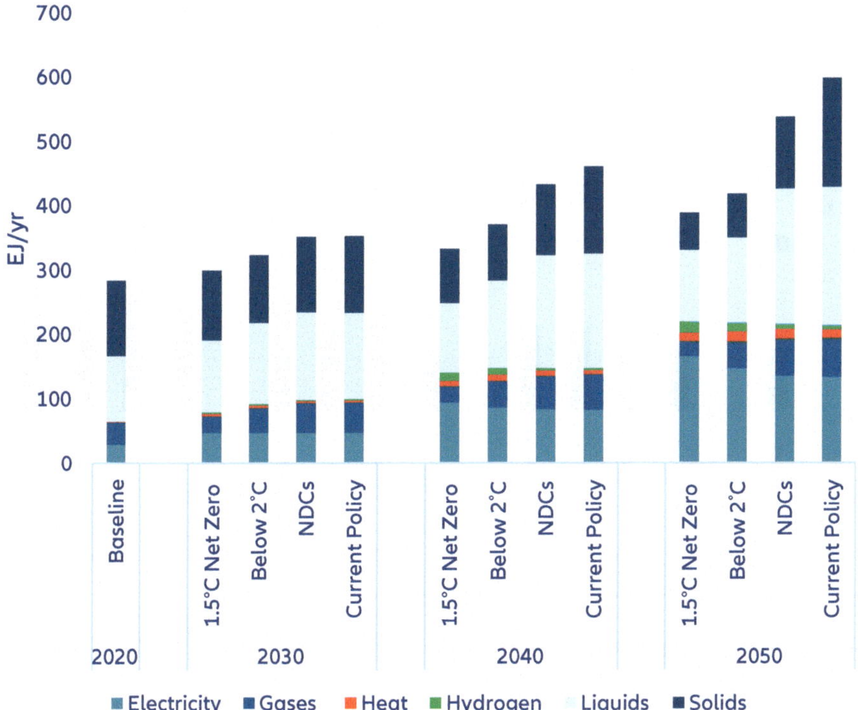

Fig. 8.3 Final energy demand by energy source, Africa's 10 largest economies. Sources: NGFS, Allianz Research. Climate scenarios described In Box 1 in the introduction to this book

In 2020, the residential and commercial sectors accounted for the highest energy consumption, which is projected to stay almost constant towards 2050 in the low-emission scenarios, while the energy demand from industry, agriculture and fishing would increase. In the transportation sector, the different pathways show noticeable differences in final energy demand. With efforts to conserve energy in transport via the electrification of the current vehicle fleet and switching to less energy-intensive modes of transportation such as rail, energy demand would not increase substantially even with an increase in population.

Figure 8.5 shows the aggregated investment needs of the largest ten African economies for different energy sources and supply-chain stages, by scenario. At USD120bn in 2030, the investment needs under the 1.5 °C scenario are sixfold the 2020 baseline. They are also 80% higher than in the 2030 current-policy scenario, the difference stemming from large additional investments in the electricity sector. By 2050, investment needs will increase to USD220bn

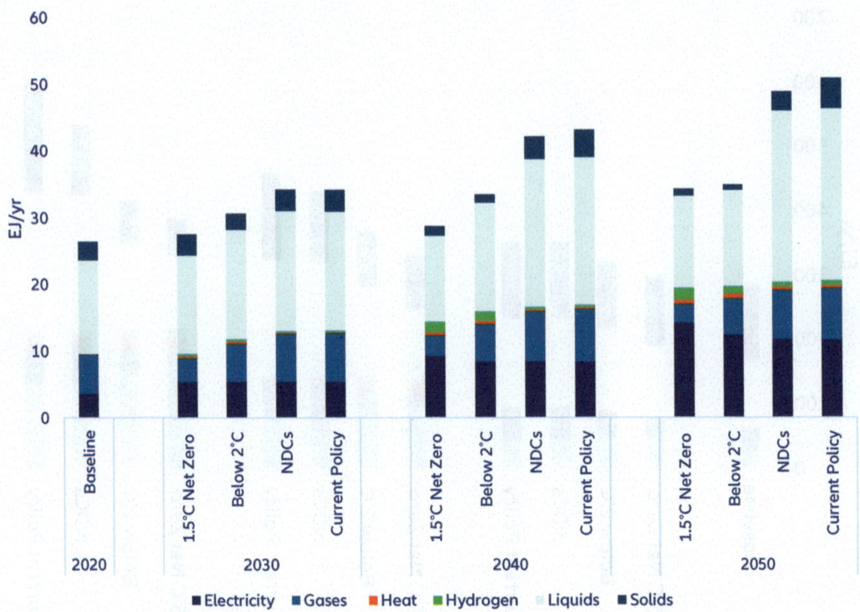

Fig. 8.4 Final energy demand by energy source, Egypt. Sources: NGFS, Allianz Research. Climate scenarios described in Box 1 in the introduction to this book

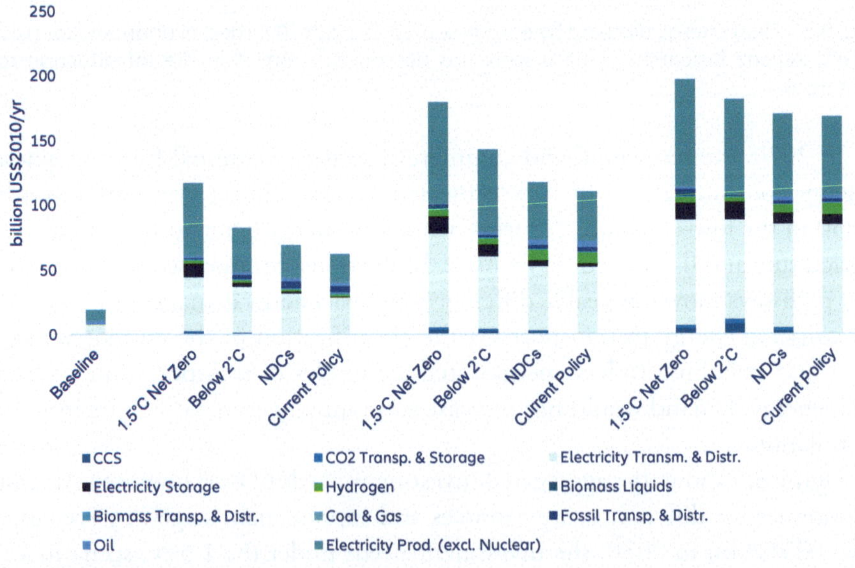

Fig. 8.5 Investment in energy supply, Africa's 10 largest economies. Source: Allianz Research. Climate scenarios described in Box 1 in the introduction to this book. Investment opportunities have been linearly interpolated between time steps. The rest of Africa has been estimated as a compound region that in the analysis features investment needs amounting to 38% of total African investments. Investment proportions in Africa as a whole are similar to those of the largest 10 economies

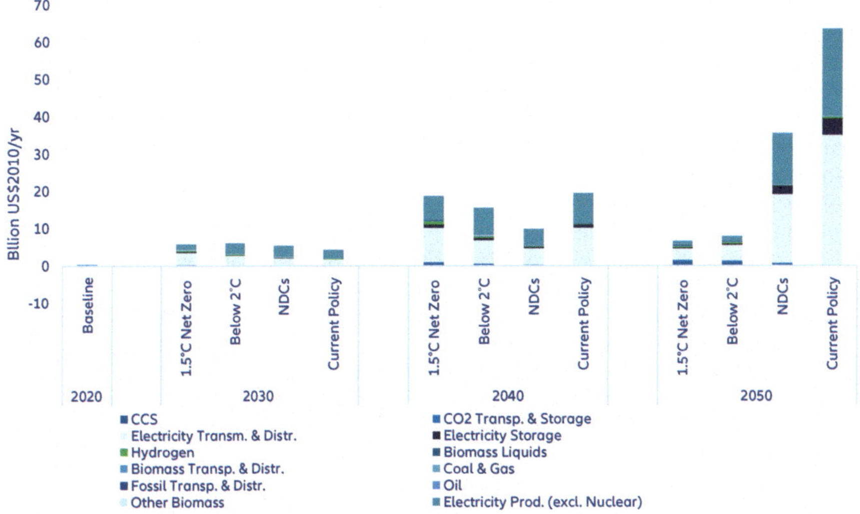

Fig. 8.6 Investment in energy supply, South Africa. Source: Allianz Research. Climate scenarios described in Box 1 in the introduction of the book

under the 1.5 °C target. For this target and Africa as a whole, investment opportunities will surpass USD200bn by 2030 and reach USD370bn per year by 2050, adding up to a total of just over USD7trn between 2020 and 2050 (see also note under Fig. 8.5).

On the supply side, Figs. 8.6 and 8.7 depict investment efforts in energy supply for South Africa and Nigeria along each of the different climate scenarios. In line with the aggregated view in Fig. 8.5, the graphs again show how potential investment strategies are dependent not only on a substantial increase in overall expenditures, particularly for electricity generation, but also on the timing of investments. Yet, the developments in individual countries vary significantly, an observation which will become even more apparent with further country examples throughout this chapter.

Mind the Gap

A transition aligned with the prospect of limiting global warming to below 1.5 °C calls for action not only far into the future, but most importantly already in the next few years. While a path consistent with a backloading of investments is possible, albeit at the expense of increasing overall costs, the general level of annual investment efforts compared to the 2020 baseline has to increase more than threefold, depending on the scenario.

The four climate scenarios show sizeable discrepancies between planned and required investment efforts. The annual investment gap between the scenario that follows current-policy commitments and the 1.5 °C-consistent pathway in Nigeria (Fig. 8.7), for example, rises in 2040 to more than double the initial current-policy investment, and ten years later to three times as much. For the aggregate of countries (Fig. 8.5), differences between the scenarios are smaller, which indicates that some countries experience a lower investment gap compared to the Paris-consistent scenarios than others.

Apart from the expansion of electricity-generation capacities, a sustainable energy transition should also include major investments in the necessary infrastructure for energy transmission and storage, since these are central to any plan for transforming the energy sector. This represents a big challenge for the selected economies, as they are far from the required national power grids or power-storage capacities. For each of the given pathways, infrastructure expansion accounts for 40–50% of total investments in the electricity sector; the absence of such infrastructure would make it difficult to attract the capital needed for investments in renewable energy projects.

When looking at Nigeria, the developments required are similar in terms of electricity generation and storage expansion. Differences are observable for the remaining investments, which concentrate on biomass, and hydrogen, and on nuclear power in Nigeria (the investments in the different categories will be further explored and decomposed in the following sections). This demonstrates that while both countries must increase their investment efforts considerably to catch up with the requirements of the 1.5 °C or even 2 °C targets, the way in which each country manages their energy transition can vary.

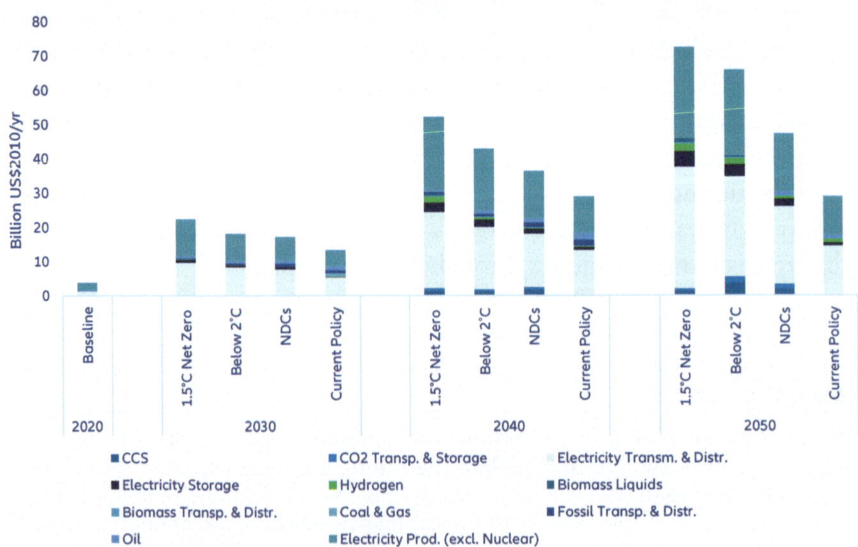

Fig. 8.7 Investment in energy supply, Nigeria. Source: Allianz Research. Climate scenarios described in Box 1 in the introduction to this book

Fossil Energy: There's Life in the Old Dog Yet

With the immense challenges accompanying the transformation of the African energy sector, the fossil industry will be subject to significant change. A shift towards renewable energy and the increasing cost of carbon emissions will put pressure on the profitability of fossil-resource extraction. On the other hand, renewable energy cannot always guarantee a sufficient supply of energy without the necessary means of storage. This will lead to lower, but persistent, extraction volumes even under climate scenarios aiming for 1.5 °C. A look at the joint projected investments in primary energy for the largest African economies in Fig. 8.8 reveals that most investment in fossil-resource extraction is directed at oil and gas. There is some remaining investment projected in the extraction of coal, which stems from the heavily coal-dependent South African economy (Fig. 8.9), but overall investments are concentrated on the other two conventional energy sources.

Figure 8.8 shows that the only scenario for which investments in fossil energy extraction decrease steadily is the 1.5 °C-consistent pathway. In all other scenarios, such investments are either constant (below-2 °C scenario) or increasing (NDC and current policy). This suggests that replacing all fossil

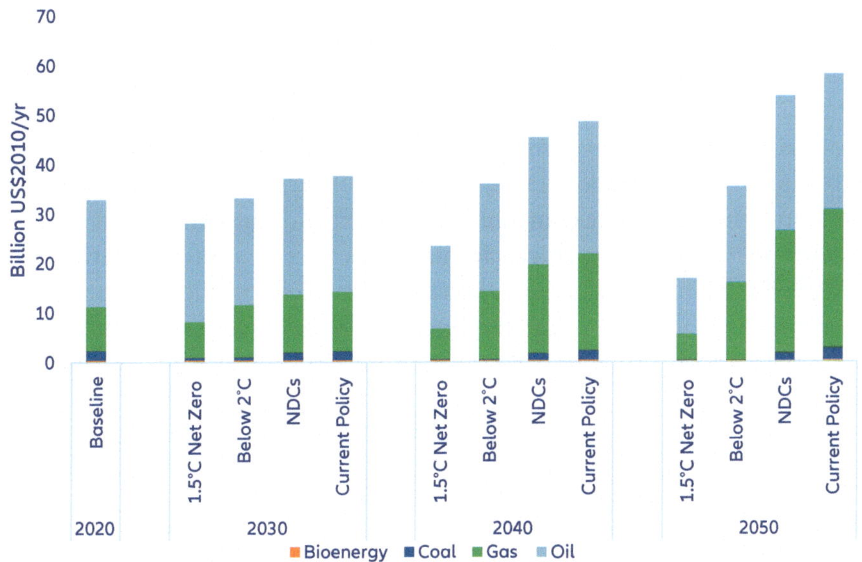

Fig. 8.8 Investment in mining, extracting and processing primary energy, Africa's 10 largest economies. Source: Allianz Research. Climate scenarios described in Box 1 in the introduction to this book

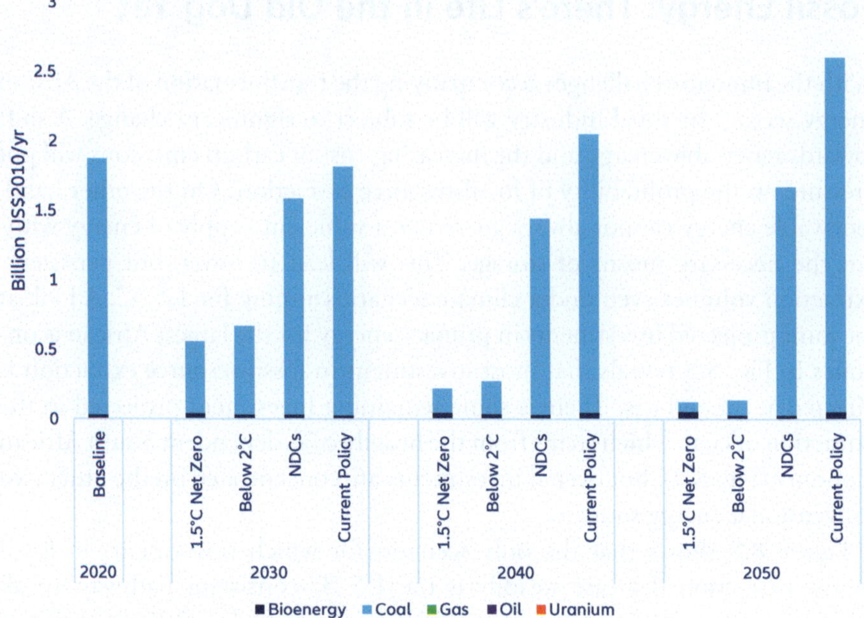

Fig. 8.9 Investment in mining, extracting and processing primary energy, South Africa. Source: Allianz Research. Climate scenarios described in Box 1 in the introduction of the book

energy might prove difficult in the near future, as renewables, due to their volatile energy yields, are not yet sufficient to cover the increase in demand.

In this case, investments in fossil-resource extraction can be partly understood as replacement investments to maintain extraction capacity without increasing overall extraction levels. Besides the investments in coal, oil and gas extraction, there are some comparatively low expenditures directed at the production of bioenergy, which can in many cases act as a direct substitute for fossil fuels. Biofuel production, however, provokes conflicts with another development goal of upmost importance: the sufficient and affordable supply of food. One country with a very strong focus on bioenergy investment is Morocco (Fig. 8.10).

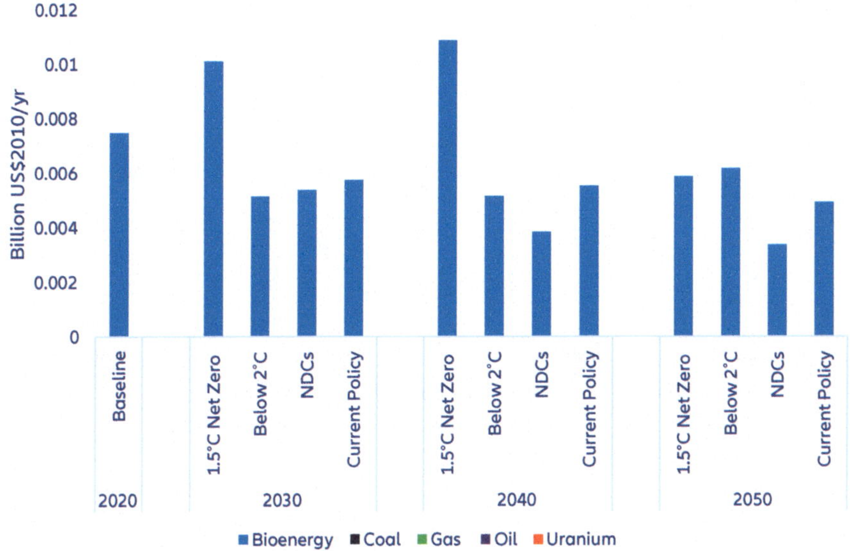

Fig. 8.10 Investment in mining, extracting and processing primary energy, Morocco. Source: Allianz Research. Climate scenarios described in Box 1 in the introduction of the book

The Future Is Electric

With renewable energy among the most cost-efficient options for electricity generation, it is not surprising that in all scenarios—even in the current-policy scenario—wind and photovoltaic dominate capacity growth in the 10 largest African economies (Fig. 8.11). This holds true for the country-specific transition pathways as well, though the importance of other power sources—particularly for other renewables—varies significantly, as Fig. 8.12 shows for the case of Tunisia. The renewable capacity stock is set to increase significantly over the years, fostered by decreasing costs for renewable energy and, depending on the scenario, substantial investments in the run-up years.

For the country aggregate, the pattern is similar, with sizable expansion of renewable-energy capacity, specifically along the pathways consistent with the Paris target of limiting global warming to well below 2 °C. Importantly, the changes currently planned and promised (NDCs) in the electricity sector will not wean African countries from fossil energy, which will still account for 20–40% of the overall capacity even by 2050 (Fig. 8.11).

Wind power facilities are unequally distributed across the continent, being tied to the geographic distribution of wind resources and policy interests. By

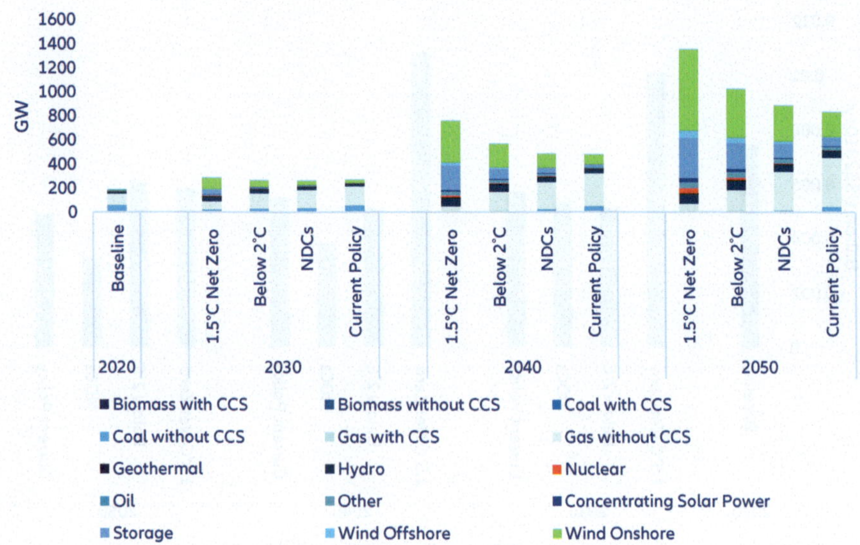

Fig. 8.11 Electricity-generation capacity by technology, Africa's 10 largest economies. Source: Allianz Research. Climate scenarios described in Box 1 in the introduction of the book

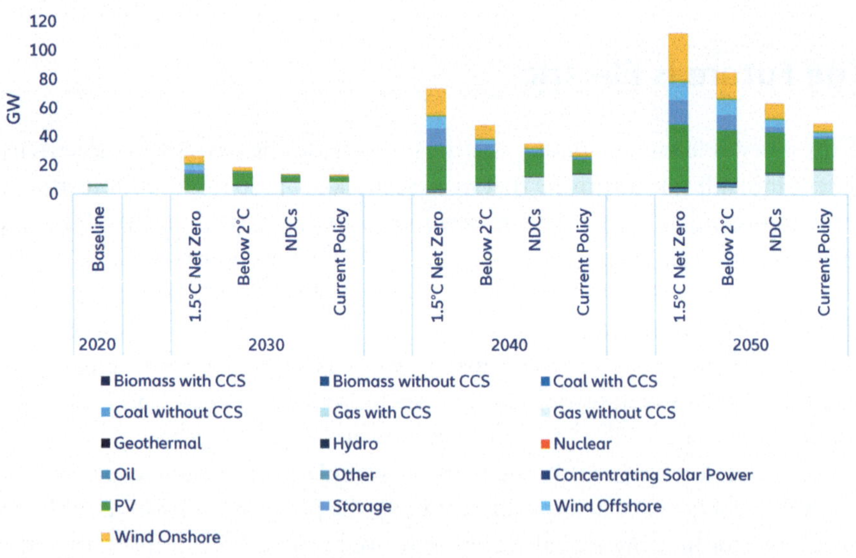

Fig. 8.12 Electricity generation capacity by technology, Tunisia. Source: Allianz Research. Climate scenarios described in Box 1 in the introduction of the book

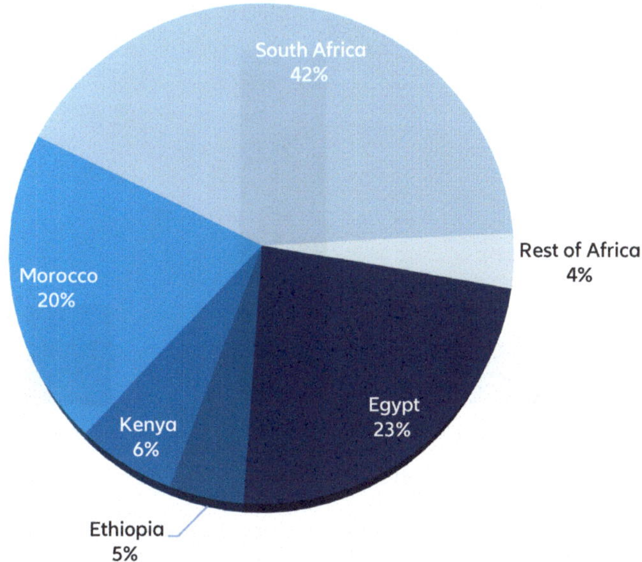

Fig. 8.13 Africa Wind Installed Capacity. Source: Allianz Research. International Renewable Energy Agency

the end of 2021, wind generation capacity was dominated by South Africa, Morocco, Egypt, Kenya, Ethiopia and Tunisia, accounting for over 96% of Africa's total wind generation capacity (Fig. 8.13). For offshore wind power generation the highest potential among the observed economies could be realized in South Africa followed by Morocco and Tunesia. Most of this potential is attributed to floating wind offshore installations (Fig. 8.14).

Figure 8.15 shows how the power capacity figures translate into investments for the aggregate of the 10 largest African economies, while Figs. 8.16 provides the example of Algeria. Figure 8.11 above gave a detailed view of investments into capacity expansion for power generation, the biggest component of energy sector investment, as well as electricity storage. Along the 1.5 °C and below-2 °C pathways, there is a clear shift in investments towards renewable energy sources such as wind, hydro or solar power. Significant investments are required for the expansion of renewable capacity in order to be consistent with the Paris goals. However, the countries show notable differences in terms of how to achieve these goals. Algeria is projected to invest relatively more in wind power. This shift towards renewables is to some extent

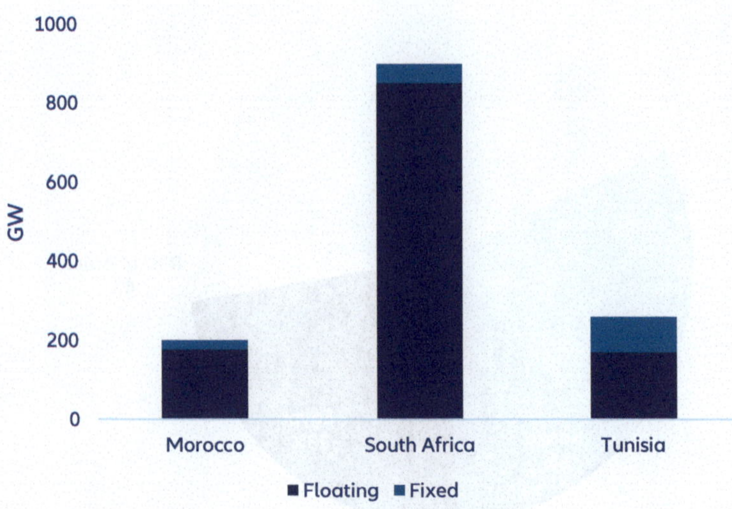

Fig. 8.14 Largest offshore wind potential in Africa. Source: Allianz Research. Climate scenarios described in Box 1 in the introduction of the book

reflected in the countries' NDC targets, but barely visible for the current-policy investments in the coming years.

Despite these efforts, investment levels along both current-policy and NDC pathways are far below what's needed to fulfill the Paris agreement's objectives, a trend consistent across the African economies studied (Fig. 8.15). The discrepancy between planned and necessary capacity investments differs by country, sometimes markedly, as shown in Algeria's case (Fig. 8.16).

A central task for African economies will lie in securing the necessary funds, both private and public, to finance the transition towards renewable electricity. The IEA's Sustainable Development Scenario projects a key role for the private sector in increasing power-generation capacity, with investments projected to rise by 400%. To achieve this, renewable investment projects must become far more attractive for private investors, with governments prioritizing the provision of a reliable financing infrastructure, well-designed competitive procurement programs for power-generation projects, and a comprehensive and credible strategy to increase clean-energy deployment.

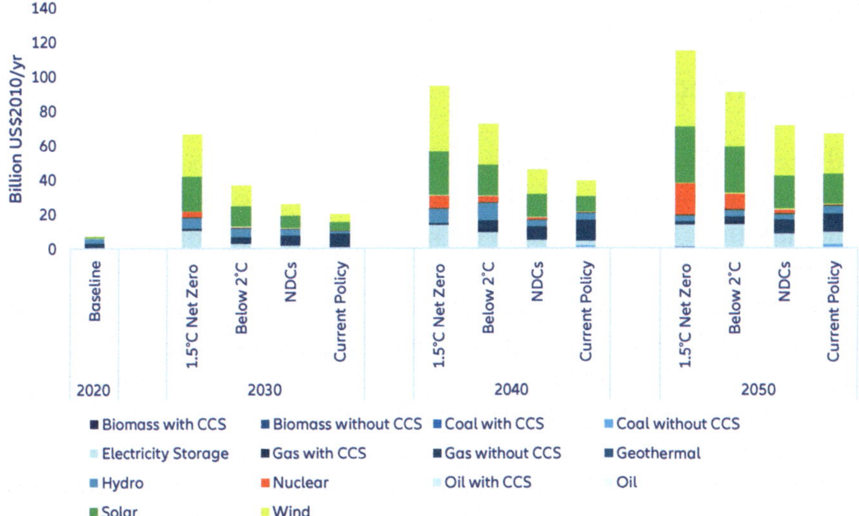

Fig. 8.15 Investment in electricity supply, Africa's 10 largest economies. Source: Allianz Research. Climate scenarios described in Box 1 in the introduction of the book

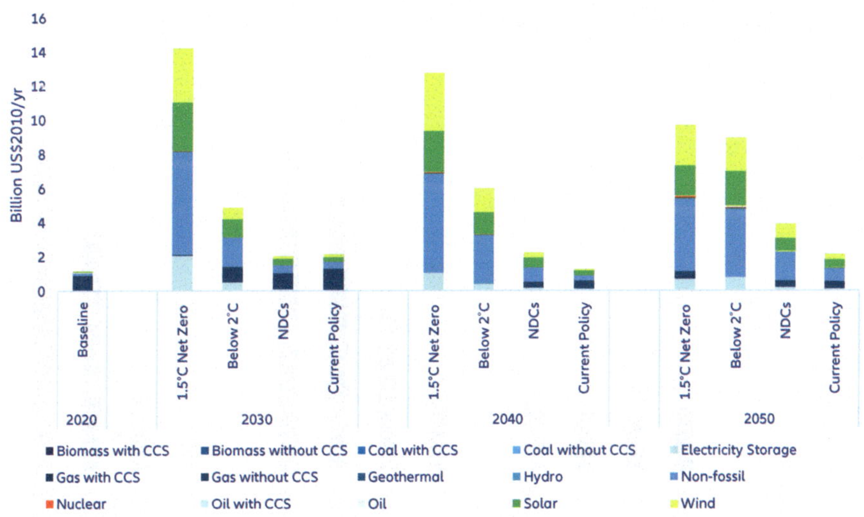

Fig. 8.16 Investment in electricity supply, Algeria. Source: Allianz Research. Climate scenarios described in Box 1 in the introduction of the book

Hydrogen

Fueling Change

With its vast low-cost renewable resources, the African continent has the potential to become one of the main exporters of low-carbon hydrogen and, at the same time, leapfrog towards the world's first hydrogen-based economies and societies. The realization of this future is envisioned in the African Hydrogen Partnership's operational plans for green hydrogen. Building on its large potential for solar and onshore wind energy, it could be an early mover in a quickly growing hydrogen market. The business opportunities opening up are very attractive and can be highly profitable in the mid and long term.

Figure 8.17 depicts investment in hydrogen-based energy supplies using different feedstocks for the 10 largest African economies. Clearly, hydrogen plays an important role in any climate scenario, with investment needs growing from USD2–3bn in 2030 to USD4–9bn in 2050. In less ambitious climate scenarios, hydrogen might even see higher investment levels in the long run, as hydrogen-based energy supplies are set to play a specific but important role in the African energy-supply mix (see also Fig. 8.3). The production

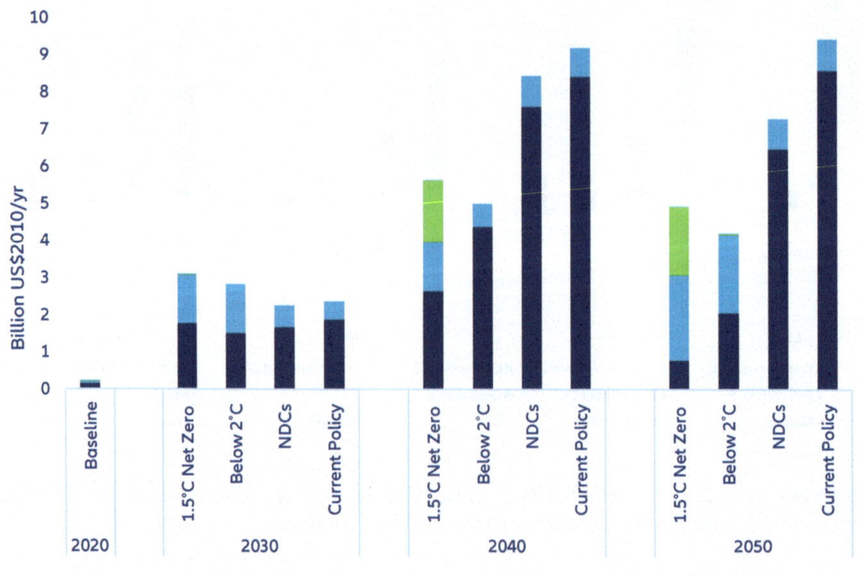

Fig. 8.17 Investment in hydrogen infrastructure, Africa's 10 largest economies. Source: Allianz Research. Climate scenarios described in Box 1 in the introduction to this book

process varies markedly among the scenarios, as the 1.5 °C scenario shifts towards processes that produce hydrogen from renewable sources while in the NDC or current policy scenarios, production continues to rely on fossil energy carriers.

In addition to using hydrogen in the domestic energy mix, favorable production conditions offer an enviable chance for African producers to establish long-term supply relations with the competitive European and Asian markets.

The African hydrogen strategies announced include the establishment of so-called "landing-zones". These are regions with favorable conditions for large-scale hydrogen projects that will serve as the basis for an expansion of such projects across other regions in Africa, and especially central Africa. One of these landing zones is Nigeria, which is already a major producer of fossil fuels. Thanks to its favorable location and coastal connections, the country is well positioned to act as a central hub for pipelines, as well as port and road traffic. At the moment, Nigeria uses hydrogen extracted from natural gas and coal, mainly for oil refining. Moving towards the production of green hydrogen would be an important part of the country's transition to clean energy, and would provide an attractive opportunity to continue to use infrastructure made redundant by the shrinkage of oil and gas sector.

Nigeria's future investments in hydrogen-based energy are set to be significantly higher than in most other major African economies. Especially in the 1.5 °C net zero scenario, there is no way around extensive investments in Nigerian hydrogen, especially from renewable electricity sources, with the associated investment demand already reaching USD2bn by 2040 (Fig. 8.18).

Ghana, though not among the top 10 African ranked GDPs, is another designated landing zone and set to act as one of the main hubs for clean hydrogen in Western Africa. The country's renewable energy potential from solar, wind, biomass, and hydropower is huge. With demand for electricity increasing by around 10% a year, domestic supply is already struggling to keep pace. This trend opens the door for renewables to step in and take a more dominant role in the energy mix. At present, Ghana covers the largest parts of its primary energy needs with imported oil and local biomass resources. According to the International Energy Agency, it is on track to reach universal access to electricity for its citizens by 2030. The only way for the country to fulfill its ambitious climate mandate is to exploit its abundant renewable energy resources and include green hydrogen in its energy mix. This creates a need for substantial financial investment and local policy support. As a result, Ghana's investment patterns will need to be very similar to those of Nigeria, though on a much smaller scale (Fig. 8.19).

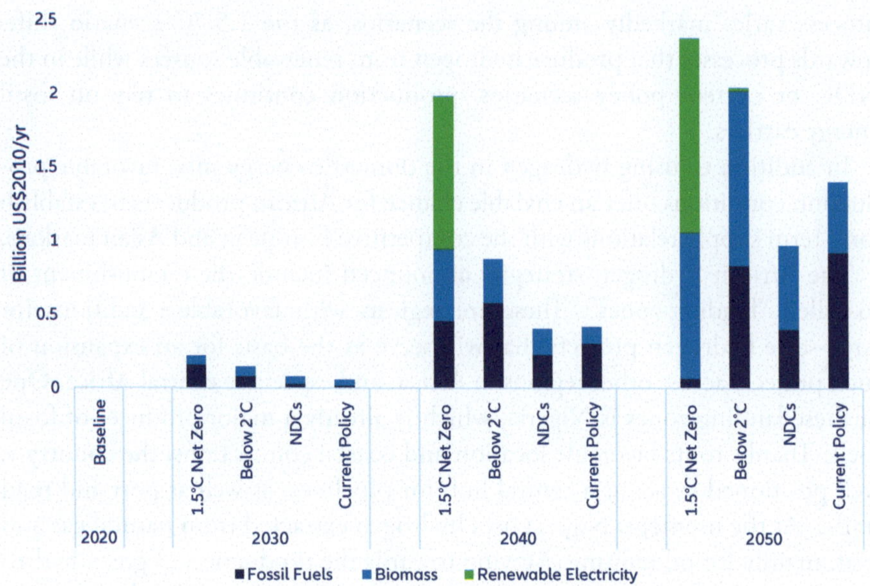

Fig. 8.18 Investment in hydrogen infrastructure, Nigeria. Source: Allianz Research. Climate scenarios described in Box 1 in the introduction of the book

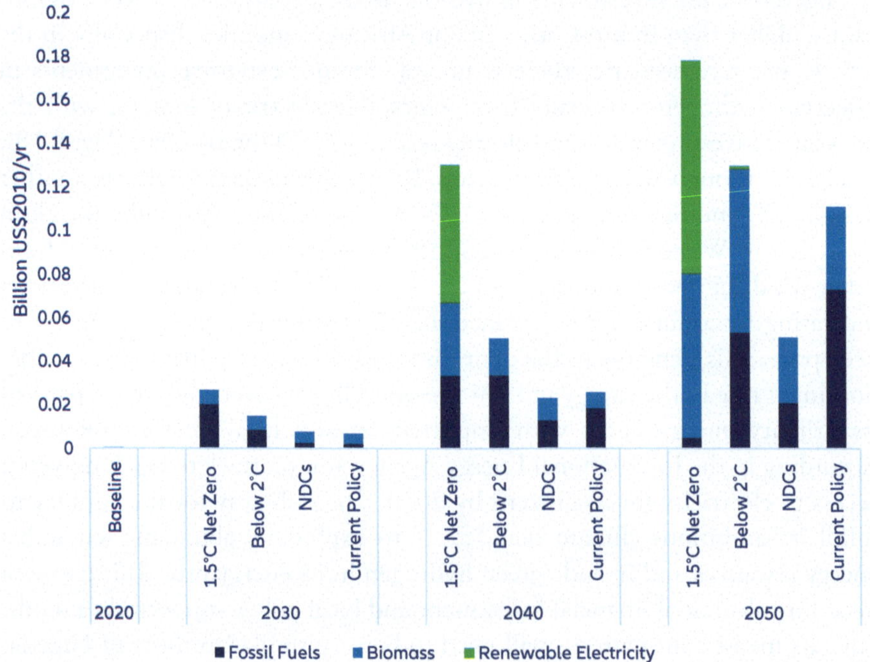

Fig. 8.19 Investment in hydrogen infrastructure, Ghana. Source: Allianz Research. Climate scenarios described in Box 1 in the introduction of the book

Conclusions

The transformation of Africa's energy sector is unfolding at a critical time, offering a unique opportunity to bypass traditional development models and move directly towards a more sustainable future. This shift carries the potential for wide-ranging societal benefits, including alleviating poverty, enhancing labor productivity, fostering agricultural growth, and improving human capital across the continent.

Africa's future energy landscape is poised for a revolution, led by a significant drive towards renewable sources like wind and photovoltaic energy. This momentum is mirrored across various African countries, each embracing unique renewable strategies, as exemplified by Tunisia's wind power expansion and Morocco's focus on bioenergy.

While the shift towards renewables is pronounced, fossil energy will continue to have a role, comprising 20–40% of the overall electricity generation capacity by 2050. This underlines the hurdles for fully abandoning conventional energy sources. Furthermore, current investment plans, though substantial, still lag behind the targets set by the Paris Agreement, demanding a unified effort from the private and public sectors.

The rise of hydrogen adds a new dimension to Africa's energy future, positioning the continent as a potential global leader in low-carbon hydrogen production. This potential growth calls for expansive investments and supportive local regulations.

However, the challenge of financing this transformation is immense, since the investments required dwarf the local resources available. The added complexities of political instability and weak legal frameworks can hinder access to necessary foreign capital.

A synergistic approach is needed to fuel the African energy transition, involving collaboration between African developing economies, advanced global economies, and private investors. Public-private partnerships are seen as a key mechanism to drive this change. The use of blended finance vehicles and support from multilateral lenders could mitigate investment risks and pave the way for more feasible projects.

As the demand for energy in Africa continues to grow, spurred by economic development and population growth, a swift, well-orchestrated response is crucial. The path forward involves reinforcing institutions, improving political governance, modernizing policy frameworks, and utilizing blended finance. Well-defined green energy strategies, ambitious yet attainable goals, and prudent implementation policies will be essential to unlock private capital. Guiding principles in this process should be transparency, accountability, adaptability, and active engagement by all key stakeholders.

Particular attention must be given to the delicate balance between energy, food security (especially concerning bioenergy), and climate goals. Seizing this moment requires immediate, concerted efforts from local governments, international partners, and private investors. The plans outlined serve as a roadmap, emphasizing that the time to act is now. To realize these ambitious objectives, a sophisticated coordination of investments, policies, international cooperation, and technological innovation is vital. The unique positioning of Africa in renewable and hydrogen markets calls for a concerted effort across sectors to harness its full potential.

In summary, a successful energy transition would present Africa with an unprecedented opportunity for growth and sustainability. The challenges are indeed formidable, but the opportunities are equally promising. Africa's energy future radiates with potential.

Correction to: Introduction

Correction to: Chapter 1 in:
L. Subran, M. Zimmer, *Investing in a Changing Climate*, Professional Practice in Governance and Public Organizations, https://doi.org/10.1007/978-3-031-47172-8_1

The original version of the book was inadvertently published without the following content. This has now been corrected.

The marching orders are clear: avert the threat posed by climate change by limiting global warming to 1.5°C. Or was it to achieve carbon neutrality by 2050? Or reaching *Net Zero*, to use its meme-friendlier moniker?

As it turns out, the two goals—1.5°C maximum warming and Net Zero—are not one and the same. In fact, reaching Net Zero by 2050 will not be sufficient to stay below 1.5°C. To muddle things further, there are several Net Zeroes, some to be achieved well before 2050, and even a couple that ought to actually go *below* zero, into negative-emissions territory.

But if one could be excused for feeling somewhat mystified by the profusion of goals, the confusion only grows worse when it comes to how to reach them. The road to Net Zero—*any* net zero—is obscured by a thicket of subsidies, bans, emissions pricing, caps, offsets, levies, international accords, national goals, behavioural changes, and a bewildering array of technological fixes and geoengineering proposals, most of which are yet to be developed.

The one thing that is indisputably clear is that getting there will be costly. But, also indisputably clear, doing nothing would be far, far costlier.

The updated version of this chapter can be found at https://doi.org/10.1007/978-3-031-47172-8_1

Governments, as well as private and institutional investors, will therefore need to be very selective when it comes to deciding where to put their money—particularly now, when state and private purses alike are very tight in the wake of the covid pandemic, the energy crisis, ballooning public debt, high interest rates, and galloping inflation. Climate action, clearly, must not only be effective, but also efficient, if only to make it easier to share the burden.

Unfortunately, when sifting through the masses of data, reports, and hype, it is very difficult to find out what really merits further attention and, most of all, where the funds should go. A thorough review of the literature more often than not comes up empty in terms of clear, credible advice in this regard.

This book aims to correct that.

The manufacturer's authorised representative in the EU is Springer
Nature Customer Service Centre GmbH, Europaplatz 3, 69115 Heidelberg,
Germany. If you have any concerns regarding our products, please
contact ProductSafety@springernature.com

Printed and bound by CPI Group (UK) Ltd, Croydon, CR0 4YY

01/07/2026

02153167-0001